Analyses in Behavioral Ecology

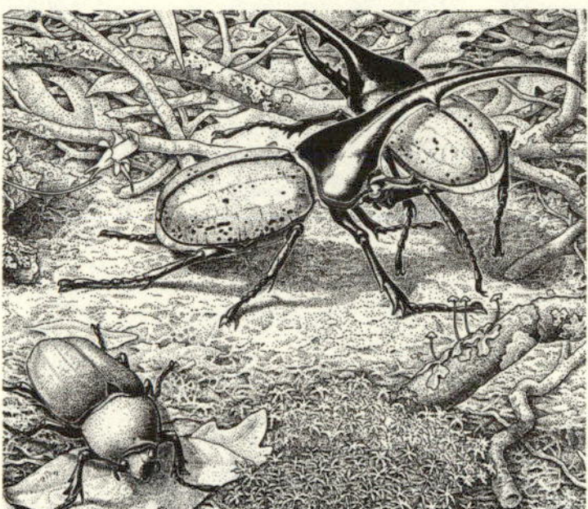

Analyses in Behavioral Ecology

A MANUAL FOR LAB AND FIELD

Luther Brown
GEORGE MASON UNIVERSITY

Jerry F. Downhower
OHIO STATE UNIVERSITY

Sinauer Associates, Inc. • Publishers
Sunderland, Massachusetts 01375

THE COVER

Hercules beetles in Venezuela; two males fight to establish dominance and obtain access to the nearby female. Original illustration by Sarah Landry from *Sociobiology: The New Synthesis* (E. O. Wilson, 1975). Courtesy of Harvard University Press.

ANALYSES IN BEHAVIORAL ECOLOGY:
A MANUAL FOR LAB AND FIELD

 For information address:
Sinauer Associates, Inc.
Sunderland, MA 01375

Original illustrations prepared for this manual by Jim F. Snyder.

Library of Congress Cataloging-in-Publication Data

Brown, Luther, 1951-
Analyses in behavioral ecology.

Includes bibliographies.
1. Animal behavior. 2. Animal ecology.
I. Downhower, Jerry F. II. Title.
QL751.B727 1987 591.5'1 87-23366
ISBN 0-87893-122-8 (pbk.)

Printed in U.S.A.

4 3 2 1

Table of Contents

REPRODUCTION

Preface

A primary difference between animals (including protists) and other organisms is that animals behave. They move about, they exhibit preferences, make choices, interact with one another, feed, mate, and fight. These activities have long been of interest to natural historians and have become the subject of intense study since Darwin's proposals of natural and sexual selection. There are many approaches to the study of behavior, ranging from the practical approach through animal husbandry to the comparative approach through psychology to the evolutionary approach through biology. Each of these involves its own sets of questions and its own methods of analysis and interpretation. Here we are interested in the evolutionary approach, an approach that draws broadly from the disciplines of ecology, ethology, evolution, population biology and genetics, all fields that are, in one sense or another, concerned with the factors that determine which organisms survive and reproduce and how they do it.

The subject matter with which we are concerned is broad. For convenience, we can divide topics into four general categories. *Sensory Capabilities* deals with the types of things that animals can perceive, the modes of perception and the limits to perceptual abilities. *Feeding Patterns* deals with the things animals eat, the ways in which they select their food, and the ways in which they interact with other foragers. *Spacing Patterns* deals with animal dispersal and dispersion, group and multispecies interactions. *Reproduction* deals with assortative mating, oviposition site choice, and sex ratios.

The following pages present some of the problems associated with these four categories of behavior. Each problem is presented so that it can be experimentally invested and analyzed. In most cases this involves the application of various statistical and graphical procedures, descriptions of which appear in the last section of the book.

Our intent when writing these investigations and analyses has been to develop enquiries and present appropriate analyses. Our enquiries are not classical labs: they have no "right" answer, they are successful if they generate data that can be analyzed so that specific questions can be answered. Past experience has shown that the answers themselves may differ for different times of the year, different geographical areas, different populations of animals, and so forth. We hope that each is approached as a new scientific enquiry rather than as a tried and true lab. With this in mind, we have suggested several analyses and approaches to each topic. Many of these can be greatly expanded if time and abilities allow, or truncated if time is more limited. Finally, we hope that the approaches described here not only expand understanding, but are fun. Animal behavior can be very exciting, and posing questions that animals can answer can be most rewarding.

LUTHER BROWN
JERRY F. DOWNHOWER

Analyses in Behavioral Ecology

Introduction

The behavior of animals is never completely self-explanatory. Questions concerning how an animal behaves and why it does so may not have obvious answers. Some behaviors may appear inexplicable from a human perspective, and even seemingly explainable behaviors raise questions. For example, honeybees are commonly seen visiting flowers. Although it may be obvious that they are collecting pollen and nectar for food, we may wonder which flowers they visit, whether they visit more than one type of flower, whether the presence of other bees affects their preferences, whether they visit each flower only once, and so forth.

The first step toward understanding animal behavior involves posing appropriate questions. We cannot begin our investigation of behavior until we know what we are inquiring about. In general, the more precise the question is the easier it will be to answer. For example, consider two questions about honeybee foraging. What is their foraging pattern? How many types of flowers do they visit? The second of these two questions is more limited and more precise. It is also easier to answer because of its precision. A series of such limited questions may ultimately reveal foraging pattern, and will do so in a manageable, answerable manner.

The second step toward understanding behavior involves collecting observations that will allow us to answer the questions we posed. In some cases this can be accomplished by watching the animal's activities without interference. Following a honeybee from flower to flower might tell us which flowers were visited. In other cases, however, we may manipulate the animal or its environment in our search for answers. If, for example, we question the effects of one bee species on the foraging of another, we may cage the bees so that we can observe them when combined and when separated. If we question the ability of honeybees to distinguish colors, we may confront them with artificial flowers of different colors and observe their preferences.

Whether our experiment involves direct manipulation of the animal or simple observation of normal activities, the data we collect are almost always quantitative. This means that we must be able to measure and record the things we study. The actual measures used are determined by the questions asked, and include counts of events, times, compass bearings, and linear or volumetric values. While it is sometimes necessary to rely on verbal or photographic descriptions, the most useful data we can obtain are numerical.

The third step toward understanding behavior involves analyzing the data to provide answers to the question we posed. Occasionally, the answers we seek will be so clear that simple observation is adequate. These occasions are rare, however. Much more often we find that data are variable. Individuals differ in their behaviors. Single individuals differ in their behavior at different times or under different conditions. Populations differ from one another. Underlying patterns may be obscured by this variability, and data may require analysis before answers become clear.

Each field of biology has its own analytical tools used to answer its own questions. The study of cell structure would not be the same without microscopes. The study of molecular genetics would be very different without tools to identify DNA and RNA sequences. Analytical tools at our disposal include various types of visual summaries, such as figures (i.e., graphs) and tables, and, most importantly, various statistical analyses. These statistical analyses are particularly important, since they allow us to provide concise answers to our questions, sometimes by succinctly describing the data we collect and sometimes by allowing us to discriminate between different possible answers. For example, observation of several foraging bees may lead us to conclude that each bee visits a different number of flowers. Statistical analyses of the data may summarize the number of flowers visited in an average value (the mean), and indicate the degree of variation among bees in a standard fashion (the standard deviation). Further analyses might lead us to conclude that the bees favored certain flower types and avoided others, or that one species of bee visited fewer flowers on average than did another species. While statistical analyses require algebraic calculations, pocket calculators and computers have made many of them quick and simple. They are valuable tools that are commonly used to answer behavioral questions.

The final step toward understanding a behavior involves interpreting the answers to our questions. This step involves deciding what our answers mean. It is only moderately interesting, for example, to find that honeybees visit an average of 3.6 flower species on each foraging trip. It is much more interesting to know how that compares to the number of flower species they encounter on a foraging trip, or how it compares to what other bee species are doing. Interpreting the answer of 3.6 involves understanding why that value (and not some other value) is important to the bee. Quite commonly, interpretation will also raise additional questions that need further observation and analysis to answer them. This sometimes leads researchers into long-term studies of multiple questions about a single species, and sometimes leads them to long-term studies of a single question in multiple species. In either case, the process of question-and-answer captures the imagination and stimulates the activities of all those who seek insight into the behavior of animals.

What follows in this text is a series of questions and suggested ways to find and interpret answers. We hope that your imaginations are stimulated and your activities are directed in a rewarding fashion. And we hope that you learn something about both the behavior of animals and the asking of questions.

1. Mechanistic Approaches to Behavior: Simple Orientation Movements

ABSTRACT

Kineses, taxes, and light reactions are examined in several invertebrates. These simple orientation movements are used to evaluate animals as mechanisms, and to develop simple simulation models of animal responses to their environment.

INTRODUCTION

The behavior of animals is generally a coordinated response to their internal or external environment. Individual behaviors are more or less predictable, given knowledge of the factors that influence them. The complexity of the behavior pattern and its predictability are usually inversely correlated. A fly, for example, may extend its proboscis every time it tastes sugar. The same fly may or may not mate every time it sees a fly of the opposite sex. The control of drinking sugar solution is obviously more rigid and simpler than that of mate choice.

The predictability of behavior allows the formulation of models for its control. Such models are analogies of the neuronal and hormonal processes that control behaviors, and may provide insight into how animals coordinate their activities and make choices. One modeling approach that has developed recently involves the use of computer simulations to deal with questions as diverse as how female weevils choose oviposition sites, how female fish choose mates, and how butterflies decide which way and how far to fly. Such models are important because they can generate predictions that can be tested by observing the live animal, thus discriminating among multiple explanations for the behavior. The figure on page 4 gives a flow chart for a model of mate choice in the fish *Cottus bairdi*. T_1 and T_2 refer to the sizes of the two neighboring nest sites. M_1 and M_2 refer to the lengths of the males at each site. Various hypotheses about female discrimination of male size were evaluated by changing the decision question from, "Is $M_2 > M_1$?" to, "Is $M_2 > M_1 + x$?", where x varied between 1 and 10 mm. Comparison of the modeled patterns with the mating patterns in a wild population suggested that the most accurate decision question was, "Is M_2 greater than or equal to M_1?"

In this exercise we want to examine some very simple animal responses to stimulation in an attempt to understand how external stimuli can be used to orient an animal's body. The goal of the analysis is to interpret orientation responses mechanistically and develop models of these simple behaviors.

TAXES, KINESES, AND LIGHT REACTIONS

Stimulation of an organism can produce directed or undirected responses. Undirected responses are termed **kineses**. These responses do not result in changes of an organism's orientation, but do result in changes in locomotion. For example, an organism may increase its speed of locomotion in response to stimulation but not change its direction of movement. *Paramecium* provides a good example of kinetic responses to light: members of these species increase their rate

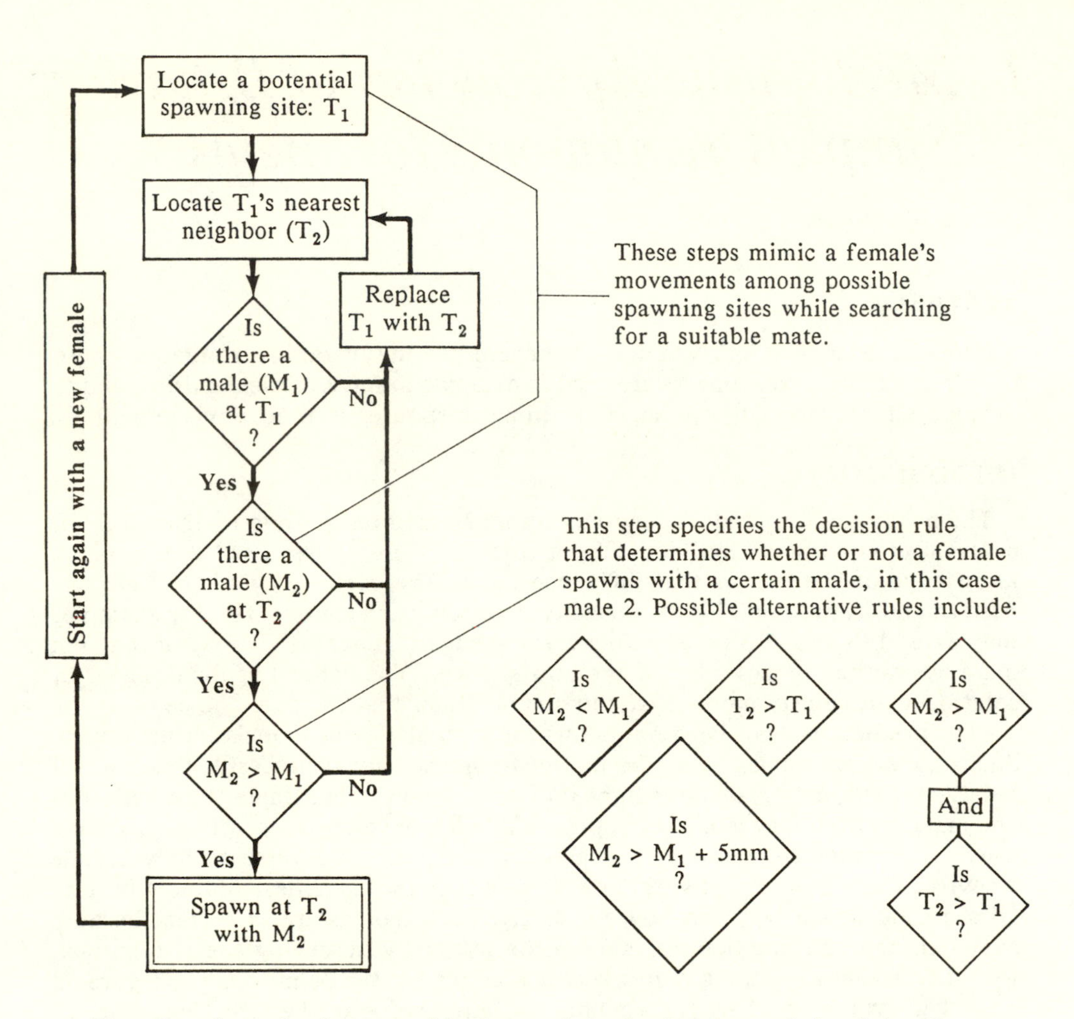

Flow chart for a simulation model of mate choice by female mottled sculpins (Cottus bairdi). *See text for description of terms. From Brown and Downhower, 1983.*

of turning when exposed to progressively higher light intensities; they do not change their overall direction of motion. Two types of kinetic responses may be differentiated from each other, and are identified in Table 1.

Directed responses occur when an animal orients its body or path of motion with respect to the direction of the stimulus. There are a number of simple directed responses that have been named. **Taxes** are directed responses in which the animal orients its body so that it points (or moves) towards (positive taxis) or away from (negative taxis) the stimulus source. Three types of tactic movements are identified in Table 1. Alternatively, animals may orient their bodies at some fixed or temporary angle to the stimulus source. If orientation is such that the animal keeps the stimulus above (or below) its body, the response is termed a dorsal (or ventral) **light reaction**. If orientation involves maintenance of a temporarily constant angle of motion with respect to the stimulus source, the response is termed a **compass movement**. Notice that taxes, light reactions, and

Table 1. Some of the categories of simple tactic, kinetic and transverse orientations proposed by Fraenkel and Gunn (1961).

GENERAL DESCRIPTIONS	FORM OF REQUIRED STIMULUS	MINIMUM FORM OF RECEPTORS REQUIRED	EXAMPLES
UNDIRECTED RESPONSES			
Kineses: Undirected reactions with no orientation of the body with respect to the stimulus.			
Orthokinesis			
Speed or frequency of locomotion dependent on the intensity of stimulation.	Gradient of intensity	Single intensity receptor	*Porcellio* (woodlice)
Klinokineses			
Frequency or degree of turning per unit time dependent on the intensity of stimulation.	Gradient of intensity	Single intensity receptor	*Paramecium*
DIRECTED RESPONSES			
Taxes: Directed responses in which the body is oriented in line with the stimulus source and movement is either towards (positive) or away from (negative) it.			
Klinotaxis			
Orientation indirect, by regularly alternating lateral deviations of the body and comparison of intensities of stimulation on each side.	Beam or gradient	Single intensity receptor	Diptera larvae, *Euglena*
Tropotaxis			
Orientation direct, by turning to the more or less stimulated side (i.e., simultaneous comparisons of intensities on both sides).	Beam or gradient	Paired intensity receptors	Woodlice, moths, caterpillars
Transverse Orientations: Directed responses in which the body is oriented at some angle with respect to the direction of the stimulus.			
Compass reactions			
Movement at a temporarily fixed angle to the stimulus.	Beam	Several receptors aimed in several directions	Ants, bees, caterpillars
Light reactions			
Orientation so that the source of light is kept above (dorsal) or below (ventral) the body	Beam	Paired intensity receptors	Brine shrimp, *Daphnia*, some fishes

compass reactions differ only in the angle at which the body is held with respect to the stimulus: 180° or 0° for taxes, 90° or 270° for light reactions, and any temporary angle for compass reactions.

THE STUDY ORGANISMS

Fly larvae (Insecta: Diptera) typically provide a clear example of negative phototaxis. Their photoreceptors are located in the tissues on either side of the condylar spine of the pharyngeal skeleton (in the head) and are alternately exposed to light and protected from light as the larva crawls forward, extending and retracting its head. Each head extension is directed to the left or right, so photoreceptors are being exposed on alternate sides of the organism. The stimulus detected during these head movements is increased light intensity; the corresponding response is movement of the head away from the stimulus (figure below).

Isopods (Crustacea: Isopoda) provide more variable examples of phototaxis and kineses. The photoreceptors are clearly visible ocelli, located on each side of the head. The intensity of light at each of these receptor sites is simply compared, and the animal turns towards or away from the more simulated side. Individual

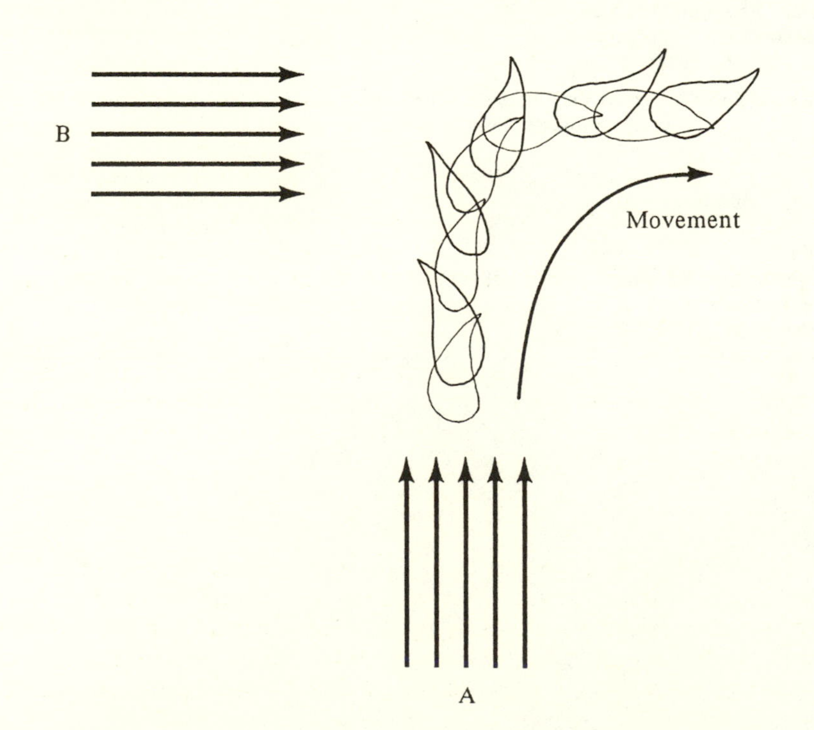

Klinotaxis in fly larvae. When light shines from "A" the larva crawls away from the source by balancing intensity on each side of the body. When light shines from "B" the larva turns away from the more stimulated side until both sides are equally illuminated.

animals seem to be either positively (rarely) or negatively (much more common) phototactic. Orthokinetic responses are given to humidity: movement increases as humidity decreases.

Brine shrimp (*Artemia*) exhibit pronounced ventral light reactions. Their photoreceptors are clearly visible compound eyes located on the head. Shrimp swim so that their ventral surfaces face the light source.

Caterpillars of many species (e.g. salt-marsh caterpillars, *Estigmeae acraea* and wooly bears, *Isia isabella*) have long been known to cross roads at right angles to the road's long axis. Various explanations have been proposed for this, the most intriguing of which is that they possess infrared radiation receptors and can orient tropotactically in response to short-wave radiation gradients.

METHODS

Examine the responses of fly larvae to a beam of light from a flashlight. Use two intersecting beams as vectors. Vary vector length by increasing or decreasing light intensity (move lights towards or away from the animal). Demonstrate that the larva responds to the vector sum. Demonstrate that the larva orients klinotactically by holding the light directly above the larva and allowing light to fall on the animal only when its head is extended to the right side.

Repeat these experiments using isopods. Since these will generally move very quickly, it may be advisable to allow them to crawl over the table while tracing their path with chalk.

Cover one of the isopod's photoreceptors with a very small dab of india ink. Illuminate the animal from above and demonstrate that it orients tropotactically.

Place a damp disc of filter paper in a petri dish and line a second dish with a dry disc. Add several grains of dessicant to the dry dish, and place ten isopods in each container. After 10 minutes, count the number of active animals in each dish. Pool the results from at least 10 replicated experiments, and demonstrate orthokinesis.

Place several adult brine shrimp in a beaker of salt water supported on a ring stand so that a flashlight can be held above or beneath it. Observe the orientation of the animals when light falls on them from above, below, and to the side of the beaker.

Draw a 1.5-m diameter circle on a large sheet of white paper. Use a sheet of black paper to erect a semicircular wall along half of the perimeter of this circle. Place single caterpillars in the center of the circle, making sure that the initial orientation of the animals is haphazard. Trace the paths of the larvae when they are illuminated from above.

ANALYSES

Most of these simple experiments will result in clear responses that require limited analysis. Demonstration of orthokinesis in isopods and of tropotaxis in caterpillars may require statistical analyses. Graphic analyses will be especially important, since they can show the orientation patterns clearly, and since they can be used to represent flow charts for models of the various movements.

Suggested null hypotheses and statistical analyses include:

a. Mean (or median) numbers of active isopods were equal in moist and dry environments: Student's t test (or Mann-Whitney U test).
b. Crawling directions for caterpillars were randomly distributed: Rayleigh test.

Suggested graphical analyses include:

a. Illustrations of typical response patterns for each condition (see especially Fraenkel and Gunn, 1961, for examples of this).
b. A bar graph indicating percent active isopods in moist and dry environments.
c. Flow charts for models of the control systems of the orientation behaviors studied.

INTERPRETATION

This set of simple experiments is designed to stimulate your thoughts about the ways that animals respond to stimuli and direct their movements. Think about the following questions when you interpret your analyses. What can you tell about the way an animal perceives ("sees") its world by its responses to simple stimulus situations like a directed beam of light? Does the fact that you can propose a simple model or computer simulation of a response tell you anything about the question of animal awareness? Why do you suppose that there is so much diversity in the ways that animals detect directional components of stimuli? What selective forces might favor each type of orientation response (i.e., what conditions might favor klinotaxis over tropotaxis?).

SUGGESTED REFERENCES

Braitenberg, V. 1984. *Vehicles: Experiments in Synthetic Psychology*. MIT Press, Cambridge, MA.

Brown, L. and J.F. Downhower. 1983. Constraints on female choice in the mottled sculpin. In *Social Behavior of Female Vertebrates*. S. Wasser (ed.). Academic Press, New York.

Carthy, J.D. 1958. *An Introduction to the Behavior of Invertebrates*. Allen & Unwin, London.

Carthy, J.D. and G.E. Newell. 1968. *Invertebrate Receptors*. Academic Press, London.

Dethier, V.G. 1963. *The Physiology of Insect Senses*. Wiley, New York.

Fraenkel, G.S. and D.L. Gunn. 1961. *The Orientation of Animals*. Dover, New York.

Gunter, G. 1975. Observational evidence that shortwave radiation gives orientation to various insects moving across hard-surface roads. *Amer. Natur.* 109(965): 104-7.

Jones, R.E. 1977. Movement patterns and egg distribution in cabbage butterflies. *J. Anim. Ecol.* 46: 195-212.

Mast, S.O. 1938. Factors involved in the process of orientation of lower organisms in light. *Biol. Rev.* 13: 186-9,209-24.

Mitchell, R. 1975. The evolution of oviposition tactics in the bean weevil, *Callosobruchus maculatus* (F). *Ecology* 56: 696-702.

Waloff, N. 1964. The mechanisms of humidity reactions of terrestrial isopods. *J. Exp. Biol.* 18: 115-135.

Warburg, N.R. 1964. The response of isopods toward temperature, humidity and light. *Anim. Behav.* 12: 175-186.

NOTES ON THE STUDY ORGANISMS

Fly larvae are available from several biological supply houses, or can be collected from compost piles or garbage cans during warm weather. Isopods are commonly found beneath rocks, logs, etc. throughout most of North America, and are also available from commercial supply houses. Brine shrimp can be raised from eggs obtained from supply houses or pet shops, many of which also sell adult brine shrimp as food for tropical fish. Appropriate caterpillars may be collected in many areas, and some species of caterpillar are available from supply houses. Many easily maintained animals may be substituted for those suggested here and are described in detail in Fraenkel and Gunn (1961).

2. Goal Orientation by Pigeons

ABSTRACT

Homing ability of pigeons is examined by releasing birds at an unfamiliar site and analyzing the compass bearings at which they vanish over the horizon. Effects of perceived time of day on homing are examined by releasing birds that were maintained on artificial photoperiod schedules.

INTRODUCTION

Animals are frequently confronted with a situation in which they must move from one place to another. Such movements may be required during dispersal from a birthplace, during searches for mates or food, following displacement by wind or water currents, or during seasonal migrations between habitats. In some cases, these movements will involve short distances over familiar ground. In other cases, movements may involve distances of hundreds or thousands of kilometers and cover expanses of unfamiliar territory.

Many animals are capable of directing their movements toward a goal. This direction may involve something as simple as piloting, or using familiar landmarks (visual, olfactory, acoustical, etc.) to locate the desired spot. Other animals are capable of true navigation, which does not rely on familiarity with the terrain covered. Many studies have shown that various vertebrates and invertebrates can navigate using the sun or stars to indicate compass direction. Other analyses have suggested that the Earth's magnetic field, infrasound, chemical gradients, and barometric pressure all help animals find their goals.

THE STUDY ORGANISMS

Rock doves or pigeons (*Columba livia*) are well known for their navigational abilities. Feral (wild) pigeons will generally home if displaced. During the last century this ability has been exploited by fanciers and researchers, who have produced inbred strains with remarkable homing abilities. Much of our understanding of animal navigation is the result of experimental displacements of these homing pigeons. In this exercise we will displace pigeons to an unfamiliar release site and analyze their vanishing bearings after release. While the release conditions can be varied in many different ways (e.g., sunny vs. overcast weather), we will release birds under sunny skies, and will use birds that have experienced either normal or shifted photoperiods for the week prior to release.

METHODS

Pigeons used in the study can be obtained from many different sources (see Notes on the Study Organisms). Ideally, the birds should come from a university loft and be experienced in displacement and release, but this is not possible at most schools. If birds cannot be obtained from a homing pigeon loft, feral birds can be legally captured in most places, and will be suitable experimental subjects.

There are several variations of the pigeon release experiment that may be performed. In all cases, the release site should be several miles from the home of the birds, and should be wide open so that birds can be followed visually until they are as close to the horizon as possible. Birds should be released individually, and watched through binoculars until they disappear. The compass bearing at which they vanished should be recorded for each bird (see figure below).

The simplest experiment that can be performed is to release 20 or more birds in the manner described. This will provide enough compass bearings to tell whether birds show goal-oriented directionality.

If 30 birds are available, divide them into two groups one week prior to testing. Half of the birds should be maintained with a normal photoperiod. The other group of birds should be maintained with a photoperiod of normal length but having dawn delayed by 6 hours. This can be accomplished by housing the birds in a cage in a room equipped with timers that control the lights. A simpler arrangement is to house the birds in a large box or carton (e.g., the kind of box that refrigerators are shipped in) that has lights controlled by a timer. All birds should have food and fresh water continually available. Release all birds in the manner described above, alternating the normal and shifted birds at release.

Several additional experiments are possible and may be used in an expanded analysis of homing. For example, a comparison may be made between the vanishing bearings of birds that have been released from the study site previously (experienced birds) and birds released for the first time. The effects of weather may also be examined by releasing birds under sunny and overcast conditions, but we recommend reading the various works by Keeton on the effects of weather before conducting this experiment, since birds need to be trained to fly under overcast conditions.

ANALYSES

We are interested in two questions: (1) Were the vanishing bearings for either group of birds random, and (2) Were the mean bearings for the two groups different from each other?

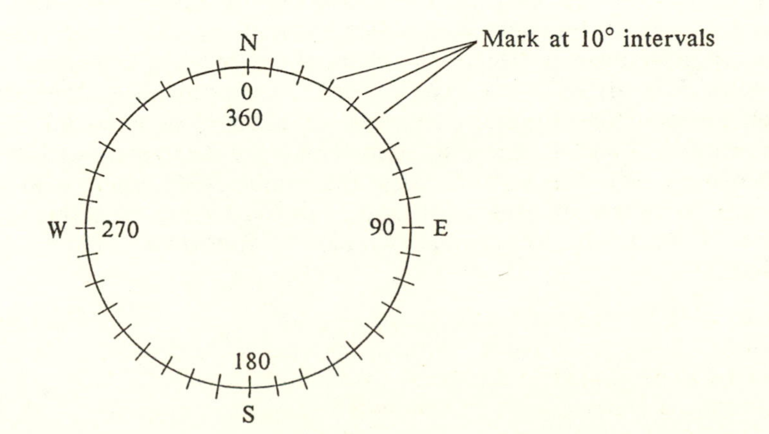

A compass rose for recording the vanishing bearings of homing pigeons.

Suggested null hypotheses and statistical analyses include:

a. Vanishing bearings for each group were random: Rayleigh test.
b. Mean vanishing bearings for the two groups were not different from each other: Circular statistic, two-sample test.

Suggested graphical analyses include:

Scatter plots of vanishing bearings, together with mean vectors.

INTERPRETATION

This set of experiments is designed to stimulate your thoughts about the ways in which animals might use the sun to help them find their way home following displacement. Think about the following questions when interpreting your analyses. Why is there so much variability in the responses of individual birds? How does the variability you observed among the pigeons compare with that you might expect if you measured visual perception abilities among your classmates? Was the mean vanishing bearing directed toward home? What was the effect of fooling the birds with a clock-shift? What does the effect of the clock-shift tell you about the way birds perceive compass directions? Do the results you found allow you to make conclusions about the importance of the sun to pigeon homing?

SUGGESTED REFERENCES

Galler, S.R., K. Schmidt-Koenig, G.J. Jacobs and R.E. Belleville (eds.). 1972. *Animal Orientation and Navigation.* NASA, SP-262, Washington, D.C.

Griffin, D.R. 1969. The physiology and geophysics of bird navigation. *Quart. Rev. Biol.* 44: 255-576.

Keeton, W.T. 1969. Orientation by pigeons: Is the sun necessary? *Science* 165: 922-928.

Keeton, W.T. 1979. Avian orientation and navigation. *Annu. Rev. Physiol.* 41: 353-366.

Zimmerman, D.R. 1979. Probing mysteries of how birds can navigate the skies. *Smithsonian* 10(3): 52-60.

NOTES ON THE STUDY ORGANISMS

Most classes do not maintain their own pigeon lofts. There are, however, a great many pigeon fanciers and racers throughout the world. Most of these private citizens are eager to teach others about their hobbies, and are usually very helpful in arranging releases of their birds during a class project. Most fanciers are members of racing clubs, and can be contacted through The American Racing Pigeon Union, Cincinnati, Ohio. The current secretary of this society can be reached through P.O. Box 95, Mainville, Ohio 45039. Alternatively, wild rock doves can be caught at their roosts and used during the experiments. However, you should check with local and state police to discover what ordinances protect the birds.

3. Color Perception in Honeybees

ABSTRACT

Various aspects of learning and color perception in honeybees are examined by conditioning bees to visit discs of colored paper. Attractiveness of different geometric shapes is evaluated by allowing bees to forage in the presence and absence of artificial nectar guides.

INTRODUCTION

Visible light includes a relatively narrow range of wavelengths in the electromagnetic spectrum. Most vertebrates and some insects see light in the range of 390 nm (violet) to 760 nm (red). This is the range of wavelengths perceived by humans. Bees, and many other insects, see those wavelengths ranging from about 300 nm (ultraviolet) to about 650 nm (orange-red). Bees thus have about the same size perceptual span, but see different wavelengths than do humans.

The simple fact that an animal can see various wavelengths of light does not necessarily mean that the organism can distinguish colors. Color vision entails the ability to discriminate between wavelengths, regardless of similarities or differences in intensities. Color vision is known among many vertebrates and invertebrates, and various lines of evidence suggest that honeybees not only possess color vision, but see their spectrum as a series of relatively distinct categories (colors), much as we do.

Because bees see different wavelengths than humans do, their colors are different from ours. The figure below presents color circles for humans and for honeybees. Primary colors are indicated by capital letters.

The differences between the perceptual abilities of bees and humans mean that

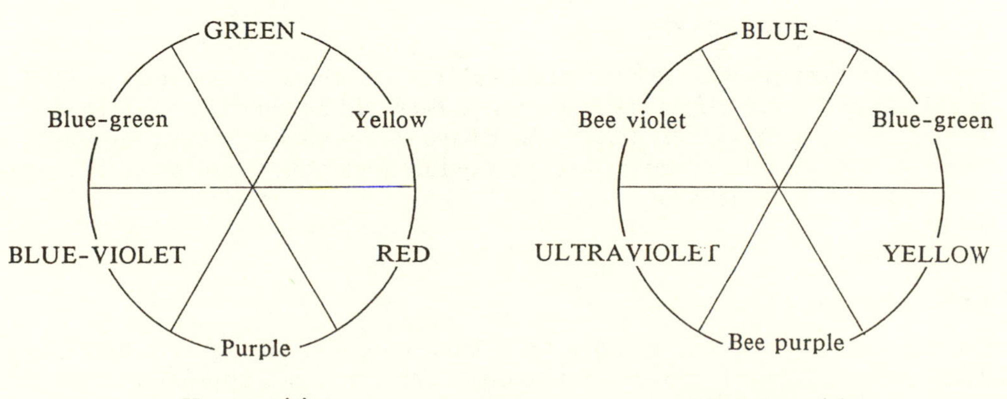

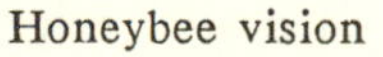

Color wheels for humans (left) and honeybees (right). Primary colors are capitalized.

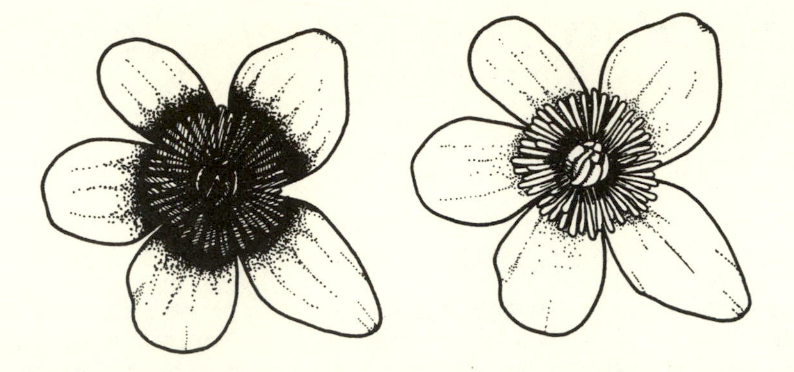

An illustration of nectar guides. Ultraviolet reflecting areas highlight the nectar- producing regions of the flower and provide a target for foraging bees. Humans cannot see these areas with their eyes. As seen by a honeybee, the flower would appear to be bluish-purple with a very dark center (left). As seen by a human, the flower would appear to be yellow and white (right).

colored objects like flowers may appear quite different to bees than they do to us. For example, some red flowers, such as poppies, are visited by bees, but bees do not see red. These same flowers reflect strongly in the ultraviolet range, suggesting that what looks red to us must look violet (bee violet) to the bees. Other flowers employ nectar guides to help bees locate nectar in the flower. Nectar guides are colored stripes or patches that point to the nectar (figure above). For example, evening primroses and cinquefoils both have pale yellow flowers that appear to be uniformly colored to our eye. In fact these flowers have strongly ultraviolet-reflecting portions of the petals that highlight the presence of nectar.

In this lab, we want (1) to demonstrate the ability of honeybees to discriminate among colors, and (2) to investigate the effects of bee nectar guides on the attractiveness of flowers to bees.

THE STUDY ORGANISMS

Our investigations will use honeybees (*Apis mellifera*) maintained in a standard beehive. On the day before the study, these bees will be conditioned to feed at small dishes of honey-water "nectar" which have been placed on blue backgrounds. The bees should thus be familiar with the feeding apparatus, the presence of food near the hive, and the color blue.

METHODS

Color Discrimination

Cut discs of colored paper to make artificial flowers of blue, gray, white, and black paper. Place each flower on a board or concrete block approximately 5 m from the hive, and place a dish of honey-water on top if it. Count the number of bees inside the disc of each flower at each 5-minute interval for 20 minutes. Maintain the level of "nectar" in each dish by adding as needed. Do not allow the bees to remove all the solution.

Repeat the procedure detailed above, but use discs of blue, yellow, red, and green.

Effects of Nectar Guides

Make two discs of blue paper by cutting around a petri dish or other suitably sized round form. Cut the discs into six equally sized pie-shaped wedges. Arrange the wedges on a yellow background, as shown in the figure below. Place the "flowers" on the ground approximately 5 m from the hive and place a dish of honey-water on top of it. Count the number of bees inside the central circle of each flower at each 5-minute interval for 20 minutes.

The experiments suggested here have used bees conditioned to feed at blue discs. The procedures can be repeated using bees conditioned to some other color (e.g., yellow) to reveal whether the bees learned to associate blue with a reward, or have some innate preference for that color. The lab can be further expanded by varying the length of the conditioning period to reveal learning rates that are color dependent.

SUGGESTED ANALYSES

We are primarily interested in discovering whether bees visit artificial flowers of different types with equal frequencies. We may also evaluate bee communication by examining the rate of visitation to sites during the course of the study. Finally, we may evaluate the effects of conditioning to different colors or for different lengths of time.

Suggested null hypotheses and statistical analyses include:

a. Bees visited the two types of "guide" flowers with equal frequency: Chi-square test; a priori hypotheses of equal frequency.
b. Bees visited each shade of gray or color with equal frequency: Chi-square test; a priori hypotheses of equal frequency

(n.b. Analyses a and b can be done on each time interval to reveal any changes in foraging during the experiment.)

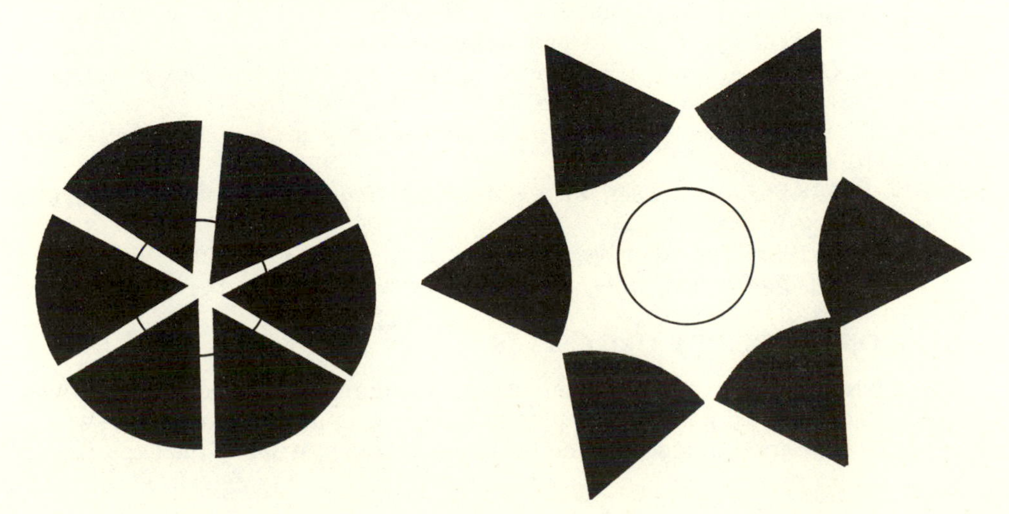

Arrangement of the "nectar guides" to be used in this experiment.

c. Recruitment rates were equivalent for all treatments: Regression analyses to examine the slopes of lines relating number of foragers to time since the beginning of the experiment.
d. The average number of foragers at each treatment was equivalent for all treatments: Analysis of variance.

Suggested graphic analyses include:

a. Scatter plots of number foragers vs. time since start for each experiment.
b. Bar graphs for number of foragers (or average number) vs. treatment.

INTERPRETATION

This set of experiments is designed to stimulate your thinking about the ways that honeybees perceive their world through vision. Think about the following questions when interpreting your analyses. Was there an effect of the artificial nectar guides? Honey reflects ultraviolet light; might this have influenced your results? Did the bees discriminate blue from shades of gray and from other colors? How might differences in the intensity of the colors confound your interpretation of the effect of wavelength? What do changes in the numbers of foragers tell you about the recruitment of workers to a food source? Did you detect any noticeable environmental effects on the attractiveness of a food source (e.g., things like shade, wind direction, proximity to the hive, etc.)?

SUGGESTED REFERENCES

Autrum, H. 1968. Color vision in man and animals. *Naturwissenschaften* 55: 586-590.

Autrum, H., and I. Thomas. 1973. Comparative physiology of color vision in animals. In Jung, ed. *Handbook of Sensory Physiology: Central Visual Information A*. Vol. VII/3. Springer-Verlag, Berlin.

Camhi, J.M. *Neuroethology: Nerve Cells and the Natural Behavior of Animals*. Sinauer, Sunderland, MA.

Hurvich, L.M. 1981. *Color Vision*. Sinauer, Sunderland, MA.

Kevan, P.G. 1983. Floral colors through the insect eye: What they are and what they mean. In C.E. Jones and R. J. Little, eds. *Handbook of Experimental Pollination Ecology*. Van Nostrand Reinhold, New York.

McCrea, K.D. and M. Levy. 1983. Photographic visualization of flower colors as perceived by honeybee pollinators. *Amer. J. Bot.* 70: 369-375.

Menzel, R. and J. Erber. 1978. Learning and memory in bees. *Sci. Amer.* 239: 102-110.

von Frisch, K. 1963. *Bees: Their Vision, Chemical Senses, and Language*. Cornell Univ. Press, Ithaca, NY.

Waser, N. M. 1983. The adaptive nature of floral traits: Ideas and evidence. In L. Real, ed. *Pollination Biology*. Academic Press, New York.

NOTES ON THE STUDY ORGANISMS

While best success is obtained when using a standard commercial beehive, wild honeybees can be easily trained to come to a feeding area (dishes of honey-water) during the lab period if conditioning begins several days prior to the experiment.

4. Perceptual Abilities of Bats

ABSTRACT

Perceptual abilities of insectivorous bats are examined by flying the bats in a plastic tunnel filled with obstacles. The effects of obstacle dispersion and size, as well as effects of bat species and sex, are examined.

INTRODUCTION

The bats (order Chiroptera) are a large, diverse, and successful group of mammals. The 853 species living today are divided taxonomically into two suborders. Megachiropterans tend to be large, and may have wingspans of almost 5 feet. They are typically fruit eaters, and have a fox-like face with well developed eyes. Most rely on vision when flying. Microchiropterans tend to be much smaller and include some of the smallest mammals. They have diversified greatly in their feeding habits and include species that feed on pollen, nectar, fruit, flesh, fish, blood, frogs, and insects. Microchiropterans rely on echolocation to detect obstacles and food items while flying.

The family Vespertilionidae has representatives throughout the world. Most of these are insectivores that use their highly refined echolocating abilities to capture insects in flight. These bats typically emit high-intensity pulses of high-frequency sound through their mouths. Sound waves that strike an object are reflected back to the bat's ears, and the bat uses information obtained from the echo to perceive the objects. The time interval between the emission of the call and the return of the echo is proportional to the distance between the bat and the object. The loudness of the echo, when coupled with information about the distance of the object, allows determination of the object's size. The clarity of the echo allows perception of the object's surface texture.

Shorter sound waves are reflected by smaller objects than are longer waves. Since many of the small Vespertilionids feed on very small insects (e.g., mosquitos), most emit very high-frequency calls. In general, there is a correlation between the size of the bat, the size of its prey, and the frequency of the call the bat typically produces. Bats of different species thus differ in their perceptual abilities. In this exercise we will examine perception in bats by flying them in a chamber filled with obstacles of known sizes. We will attempt to examine differences in the perception of different obstacle sizes and spatial patterns, and will examine differences between the sexes and species if sufficient bats are available.

THE STUDY ORGANISMS

Bats used in this exercise will be obtained from barns or house attics one or two days prior to the study. Availability will determine which species and sexes are used. Ideally, we will compare at least two species of different sizes. Likely species include the big brown bat *Eptesicus fuscus*, the eastern pipistrell *Pipistrellus subflavus*, and the little brown bat *Myotis lucifugus*.

While this analysis is admittedly time consuming and demanding, we feel it most worthwhile and rewarding. It generates abundant questions, and uses an important group of mammals that may be unfamiliar to most experimenters.

METHODS

The bats will be flown in a flyway constructed in the lab. While the exact dimensions of the flyway will be determined by the size of the lab room, it should be at least 10 m long and 2 m wide. The easiest way to build this enclosure is to suspend sheets of transparent plastic from the lab ceiling using duct tape to hold the plastic in place (see figure). Cut the plastic to make entrances at either end of the tunnel, and close these entrances by overlapping the sheets at least 2 m. The plastic should hang freely from the ceiling of the room to the floor, and continue at least 1 m along the floor as a skirt. The flyway should be arranged in the room so that students can make observations from both sides, and a "batman" or "batwoman" can walk from one end of the tunnel to the other.

Strings hanging from the ceiling of the tunnel will be obstacles to the bat's flight. These obstacles will be arranged in alternating groups of two or three strings along the length of the tunnel, thus forcing the bats to either slalom as they fly or else collide with the strings. The strings may be taped to the ceiling or hung from short wires (e.g., bent paper clips) pushed into the ceiling material. Each set of strings must be at least 1.5 m away from the next. The distance between the strings that compose a set should be approximately two bat wing-spans, or about 0.5 m, depending on the type of bat used.

Each obstacle should hang from the ceiling to within a centimeter or two of the floor. It should be weighted with a small block of plasticine clay to hold it taut. Bats should be flown under dim light, which will make the obstacles themselves invisible to humans, but collisions by the bat will be obvious since the clay block will jerk in response to the "hit." Many different sizes of obstacles can be tested; nylon fishing lines come in convenient small, medium, and large sizes (e.g., 2 lb., 10 lb., and 25 lb. test lines). Each bat should take its turn in the tunnel while it is hung with each of the obstacle types.

The data we require to answer questions about bat perception are simply the number of times each individual hit or missed each obstacle. These data are easily collected if there are many observers. If enough people participate, observers can be stationed on either side of the tunnel, in the same plane as each obstacle set. Observers on one side of the tunnel can watch the plasticine blocks and record the number of times a "hit" occurred for that set (i.e., the number of times a block jerked during the trial). Observers on the other side of the tunnel can watch the bat as it flies and count the number of times it passed through the plane of obstacles in front of them. Knowing the number of passes and the number of hits yields misses by subtraction. Each bat should be allowed to fly back and fourth through the tunnel until there are a total of at least 100 passes for the entire flight.

Most bats recognize that the walls of the tunnel wall are slippery and will not support them. They will usually land on an insect net held against the wall, and can thus be removed from the tunnel following their flight.

Bats differ greatly in their willingness to fly under these conditions. While some will fly back and forth continually until netted, others will land on any unplasticized surface, or will refuse to fly at all. Covering the ceiling of the tunnel with plastic sheets may be necessary if the lab ceiling is made of acoustical tile, or if there are any recessed light fixtures or ventilators that the bats might enter.

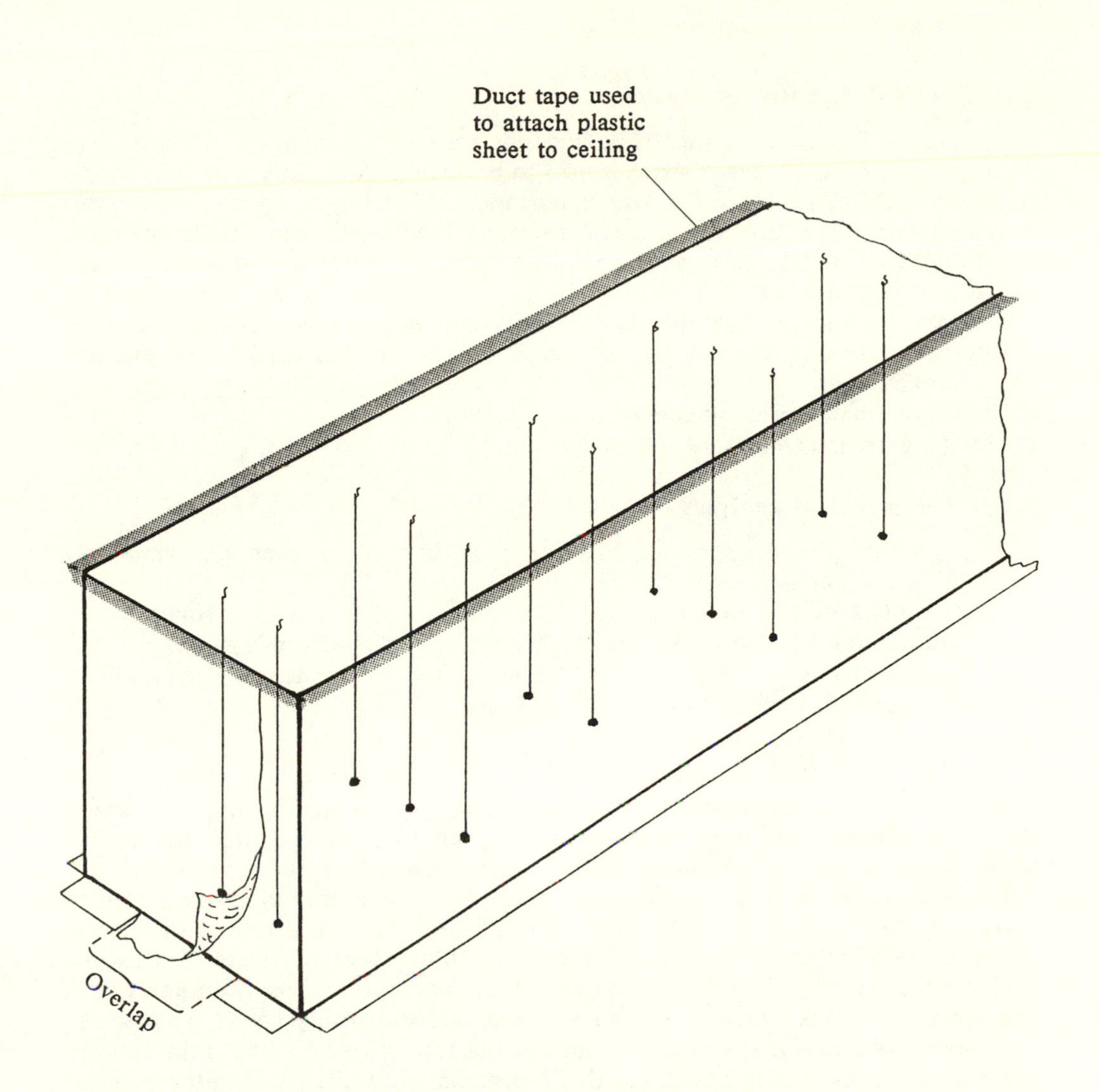

Diagram of the flyway used to test obstacle avoidance of bats. The walls of the flyway are of plastic sheeting that is suspended from the ceiling. Obstacles are arranged in planes perpendicular to the long axis of the flyway, and occur in alternating sets of two or three obstacles. Each obstacle hangs from the ceiling to the floor, where it is weighted with a small block of clay. The short skirt along the floor ensures that bats landing on the floor cannot escape.

ANALYSES

We are primarily interested in the following questions: Did individual bat performance depend on the density of obstacles (two vs. three strings)? Did individual bat performance depend on the size of obstacles (small vs. medium vs. large)? Were there differences in the performances of different bats? All of these questions can be answered through appropriate use of the Chi-square contingency table.

Suggested null hypotheses include:

a. Individual bat performance (hits and misses) was independent of obstacles dispersion (two vs. three string gates) (n.b., keep string sizes separate).
b. Individual bat performance was independent of string size (n.b., keep gate sizes separate). If this hypotheses is rejected, then contingency table analyses should continue to reveal which string sizes were equal and which were different in effects.
c. Bat performance was independent of individual bats. Rejection of this hypothesis should be followed by examination of where the differences among individuals lie.
d. Bat performance was independent of species.
e. Bat performance was independent of sex.

Suggested graphical analyses include:

a. Bar graphs of proportion of hits for each bat under each experimental condition. Alternatively,
b. Scatter plots of proportion of hits or misses for each bat as a function of obstacle diameter. Note that all of the bats can be put on a single set of axes if points for each bat are connected in some clear fashion (e.g. solid, dotted, dashed, lines for three different bats).

INTERPRETATION

This set of experiments is designed to stimulate your thoughts about the ways that bats use echolocation to perceive their environment. Think about the following questions when interpreting your analyses. Was there an effect of obstacle distribution on bat performance? Which obstacle density (2- or 3-string set) gave the bats more trouble? Do you think this effect was due to problems of perception or maneuverability? What was the effect of obstacle size on performance? Can you ascertain the limit of size perception from your analyses? Was there some size above which performance was uniformly good? What were the effects of individual differences among the test subjects? Why didn't all of the bats perform equally well? Was there as much variability among the bats as you would expect there to be in the visual abilities of your classmates? Did any of the bats show evidence of learning during the course of the experiments (e.g., can you separate any effects due to learning the flyway from effects due to obstacle size? Did any bats learn where to land in the flyway so that they didn't have to fly at all?) How do your estimates of bat perceptual ability compare with those reported by other researchers? How would you expect the feeding habits (prey, foraging habitat, foraging time) of a bat to affect its perceptual abilities?

SUGGESTED REFERENCES

Fenton, M.B. and J.H. Fullard. 1980. Moth hearing and feeding strategies of bats. *Amer. Sci.* 69: 266-275.

Gould, E. 1955. The feeding efficiency of insectivorous bats. *J. Mammal.* 36: 399-407.

Griffin, D.R. 1958. *Listening in the Dark.* Yale Univ. Press, New Haven, CT.

Neuweiler, G. 1980. How bats detect flying insects. *Physics Today* 80: 34-47.

Simmons, J.A., M.G. Fenton and M.J. O'Farrell. 1979. Echolocation and the pursuit of prey by bats. *Science* 203: 16-20.

Vaughn, T.A. 1974. *Mammalogy.* Saunders, Philadelphia.

Winsatt, W.A. 1977. *The Biology of Bats.* Academic Press, New York.

NOTES ON THE STUDY ORGANISMS

Bats can be collected from attics and barns during the summer throughout North America. Pest exterminators can usually refer collectors to people who think they have a bat "problem." Bats should be handled very gently, with leather gloves. They should be collected the day prior to the experiment, given water through an eye dropper, and released immediately after the exercise. This exercise does not harm the bats. Should bats be maintained for more than a few hours, they may be fed a mixture of low-fat cottage cheese and boiled egg yolk. Bats collected from caves in the winter will not perform well and should not be used. Note that many or all species of bats are protected in most states and collecting permits may be required of the instructor. Also note that some bats carry rabies even though they display no symptoms.

5. *Prey Location in Squirrels*

ABSTRACT

Sensory systems used by foraging squirrels are evaluated by variously camouflaging the appearance and smell of nuts.

INTRODUCTION

Squirrels are common inhabitants of the temperate forests of North America. These rodents live in fixed territories and are active more or less year-round. One characteristic of squirrels is their hoarding of food. Some species, like the red squirrel *Tamiasciurus hudsonicus* cache food (pine cones) in large piles located in the center of their territories. Other species, like the gray squirrel *Sciurus carolinensis*, scatter-hoard food by burying individual nuts for later relocation.

There are several problems associated with scatter-hoarding as a food storage system. The most obvious is that scatter-hoarded food must be relocated before it can be consumed. This problem may be especially severe for squirrels, since they bury their nuts during the early fall. Scattered nuts may then be reburied by falling leaves, and later snow. Obviously, some proportion of the buried nuts are never relocated by their buriers. These lost nuts may germinate and produce new trees, suggesting complex ecological and evolutionary relationships between nut-producing trees and their predators/dispersal agents, the squirrels.

Here we will investigate the sensory systems used by squirrels when they locate food items, and apply these findings to the relocation of hoarded food. Specifically, we will be interested in whether visual or olfactory senses are involved in food location.

METHODS

We will use unroasted peanuts as our food source. Before going to the study site, prepare several hundred "odorless nuts" by wrapping peanuts in plastic wrap and sealing each with plastic tape. Choose a site that is well forested and has abundant squirrels. Establish a baseline transect that is divided into 2-meter intervals. At each interval, establish another transect that is perpendicular to the baseline. Place the experimental nuts along these shorter perpendicular transects. Nuts should be treated in four different ways: (1) exposed unwrapped nuts (i.e., clearly visible on the surface of the leaf litter); (2) buried unwrapped nuts (i.e., covered with one or two leaves so that they are not visible); (3) exposed plastic; and (4) buried plastic. Place each nut 2 m from its neighbor and mark its position with an applicator stick stuck into the ground next to the nut. Keep all rows straight, and measure distances as accurately as possible.

Check the nuts regularly during the next several hours. When squirrels have taken approximately half of them, resample and record the number of nuts taken from each treatment.

ANALYSES

We are interested in knowing whether the squirrels found nuts in all treatments with equal frequency. If they did not, then we want to know which nuts were found most often.

Suggested null hypotheses and statistical analyses include:

An equal number of nuts were removed from each treatment: Chi-square test, a priori hypothesis of equal numbers removed.

Suggested graphical analyses include:

Bar graphs of the frequency of removal for each type of nut would provide an appropriate visual accompaniment.

INTERPRETATION

This set of experiments is designed to stimulate your thoughts about the ways that squirrels use vision and olfaction to find food. Think about the following questions when interpreting your analyses. Were there differences in the predation rates for the four types of nuts? Which nuts were found most quickly? What does this tell you about the sensory modality that first reveals prey to a squirrel predation? What does this suggest about squirrel senses? If you were a nut-producing tree, what strategies might you follow to minimize location of your nuts by predators such as squirrels?

SUGGESTED REFERENCES

Cahalane, V.H. 1942. Catching and recovery of food by the western fox squirrel. *J. Wildlife Manag.* 6: 338-352.

Nichols, J.T. 1927. Notes on the food habits of the gray squirrel. *J. Mammal.* 8: 55-57.

Stapanian, M.A. and C.C. Smith. 1978. A model for seed scatter-hoarding: Co-evolution of fox squirrels and black walnuts. *Ecology* 59: 884-896.

NOTES ON THE STUDY ORGANISMS

Virtually any species of squirrel (or other seed-eating rodent) can be used in this study.

6. Releasers and Egg Laying by Apple Maggot Flies

ABSTRACT

Stimuli that release oviposition by apple maggot flies (*Rhagoletis pomonella*) are examined by confronting the flies with models of apples.

INTRODUCTION

Many animals perform stereotyped behaviors in response to particular, often very limited, stimuli encountered in their environments. For example, male sticklebacks (*Gasterosteus aculeatus*) are fish that have typical vertebrate eyes which should allow them to see complicated forms and colors. When these males are reproductive they become territorial and very aggressive. They attack other males, and any intruding object that has a red ventral surface. These objects need not resemble other sticklebacks to elicit attack. However, they *do* need to have a red bottom. Similarly, males court females and any object that has a non-red, swollen, ventral portion, whether it looks like a fish or not. Apparently stickleback males do not evaluate all visual stimuli before attacking or courting. Rather, they respond to a very limited portion of the stimuli perceivable. That part of the total stimulus configuration that produces the behavioral response is called the **releaser** or **sign stimulus.**

The simplicity of many releasers allows them to be studied through the use of models. For example, newly-hatched herring gull chicks (*Larus argentatus*) peck at the tip of their parents' bills to stimulate the parent to regurgitate food. Several investigators have used models of gulls to show that the releaser for this behavior is the obvious red spot on the long, thin bill of the parent. (See figure below.)

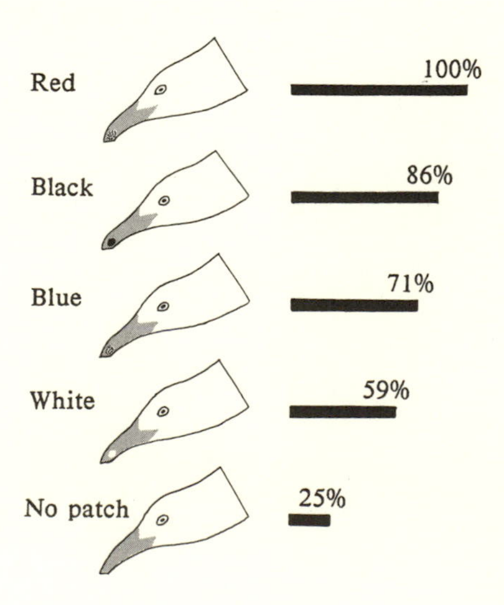

Begging response by newborn herring gull chicks to models of an adult gull head. The models were two-dimensional, and varied only in the color of the spot on the end of the bill. Responses are expressed as percentages of the response given to a red spot—the color actually found on adult gull bills. These results suggest that a red spot is the most effective releaser of begging behavior. (From Alcock, Animal Behavior, *1975.)*

Other features, such as three-dimensionality or the presence of a parental head or body, are not important.

THE STUDY ORGANISMS

In this experiment we will examine the releasers that stimulate oviposition by apple maggot flies (*Rhagoletis pomonella*). These fruit flies oviposit on developing apples throughout the summer (figure below). Their larvae then burrow throughout the flesh of the apple (they are one of the proverbial "worms" in apples). In many orchards these insects are major pests, and they must be controlled with insecticides or, increasingly, by luring them to traps with oviposition releasers.

METHODS

There are several stimuli that *Rhagoletis* might respond to while searching for an apple. Obvious clues to appleness include color, shape, size, and smell. The first three of these can be easily modeled and form the basis of our experiments. All of the experiments described here can be conducted simultaneously, and will typically yield analyzable data within a day or two.

The importance of color

Color effects are easily analyzed using rubber balls that are each approximately 10 cm in diameter. Spray paint five balls each red, yellow, green, and blue. Pass a short (ca. 25 cm) length of wire through each ball and bend it at the top to make a hanger. Completely cover each ball with "tangle-trap," which is a colorless liquid that is extremely sticky and will trap all flies touching the "apple." (Tangle-trap can be purchased from orchard, forestry, and nursery supply companies.) Hang the balls in apple trees, ideally hanging a set composed of a ball of each color in each of five different trees. Check the balls at daily intervals until approximately 20 flies have been trapped in each set, and count the number of flies on each color trap.

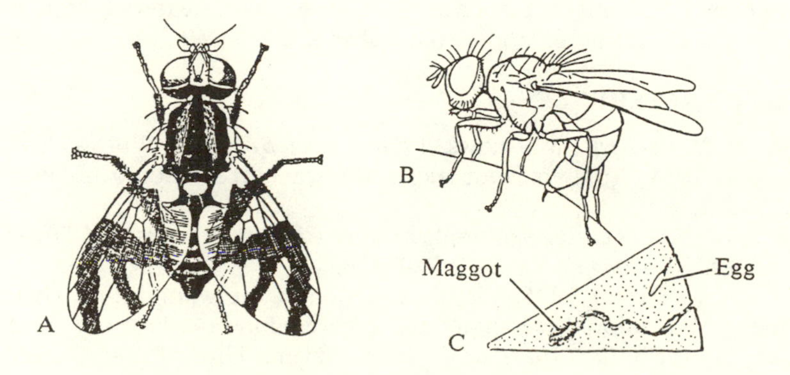

A. *An adult female apple maggot fly,* Rhagoletis pomonella. **B.** *The female punctures the skin of an apple to deposit an egg.* **C.** *A section of apple, showing an egg, and a maggot tunneling into the pulp. (Courtesy of USDA.)*

The importance of shape

The importance of geometric shape can be analyzed using styrofoam spheres, cubes, cylinders, and pyramids. These are commonly available at crafts shops, or can be cut from blocks of expanded styrene plastic. Spray five of each shape red, add hangers and tangle-trap, and repeat the experimental procedure described above.

The importance of three dimensionality can be evaluated using circles, squares, rectangles, and triangles cut either from wood or cardboard, sprayed red, and treated as described above.

The importance of size

Effects of size can be evaluated using balls of different sizes. Rubber balls of various sizes are commercially available, and many of these will provide convenient models. Use at least three sizes, ranging from 2 cm to 20 cm in diameter. Spray five of each size red, and treat as described above.

ANALYSES

In each of these experiments we are interested in knowing whether an equal number of adult flies were captured by each of the treatments. This question is easily answered using the Chi-square statistic to evaluate the null hypothesis that equal numbers were captured. Bar graphs indicating the numbers (or percentages) captured at each trap type are appropriate graphical analyses.

INTERPRETATION

This set of experiments is designed to stimulate your thoughts about the environmental cues that elicit behavioral responses from animals. Think about the following questions when interpreting your analyses. What are the most important clues used by female apple maggot flies when recognizing an apple? Are these reliable clues? If you were an apple tree, could you do anything to prevent detection? How would you design a trap to control apple maggots in an orchard? Why might selection favor a relationship between simple releasers and behavioral responses? When would selection not favor simple releasers? What is a supernormal releaser? How might responses to supernormal releasers by apple maggot flies effect the evolution of apple color, shape, and size?

SUGGESTED REFERENCES

Matthews, R.W. and J.R. Matthews. 1978. *Insect Behavior*. Wiley, New York.

Oatman, E.R. 1964. Apple maggot trap and attractant studies. *J. Econ. Entomol.* 57: 529-531.

Prokopy, R.L. 1972. Response of apple maggot fly to rectangles of different colors and shades. *Environ. Entomol.* 1: 720-726.

Riedl, H. and R. Hislop. 1985. Visual attraction of the walnut husk fly (Diptera: Tephritidae) to colored rectangles and spheres. *Environ. Entomol.* 14: 810-814.

Tinbergen, N. 1969. *The Study of Instinct*. Oxford Univ. Press, Oxford.

NOTES ON THE STUDY ORGANISMS

Apple maggot flies are common throughout the eastern United States. They are easily recognized, being about 1 cm long, and resembling a housefly that has striped wings. (See figure on page 25. A color plate appears in *A Field Guide to the Insects*, D.J. Borror and R.E. White, Houghton Mifflin, Boston, 1970.) Orchard owners should welcome this experiment, since it will remove pests from the vicinity. In other areas, the same experiments may attract other apple pests. Greenhouse whitefly (Homoptera: Aleyrodoidea) can be substituted as an experimental organism, although models for the size and shape experiments should be painted yellow, not red.

7. *Food Value and the Foraging Preferences of Squirrels*

ABSTRACT

Whether or not squirrels make choices among food items on the basis of energetic value of particular items is examined by offering individual squirrels food items that differ in size and weight.

INTRODUCTION

Squirrels of the genus *Sciurus* are scatter-hoarders. They collect items and bury them for future use. In another exercise we examined the sensory basis for the recovery of buried nuts. Here we wish to examine whether or not squirrels discriminate among food items on the basis of the quality of the food items.

There are several ways in which food items could differ in quality, but two obvious characteristics are size and weight. If food items differ in size, then larger items might be preferred by discriminating squirrels. The reasons for this follow from simple surface-to-volume relationships. A squirrel must cut through the shell of a nut in order to get at the nut meat. If the shell thickness does not increase with nut volume, then a squirrel will obtain a larger nut meat for the same amount of effort if it chooses larger nuts. In the most general sense, though, we can only *postulate* that the weight of the nut increases faster than the shell thickness, and that as a consequence a squirrel may invest less energy per calorie gained by choosing larger nuts.

Nuts of the same size may differ in weight. For example, a substantial literature documents the intensity of seed predation by insects. As an insect larva bores through the nut meat, it converts the contents of the nut into larval tissue, feces, and inevitable heat losses through respiration. Those of you who have cracked open black walnuts are well aware that a seed predator can render the contents of a seed useless to a granivore. In addition, some of you may be aware that a common and effective way to separate sound beans from infested ones is to toss the beans in a pan of water and throw out the ones that float. In short, the presence of seed predators can reduce the quality of the seed for a consumer such as a squirrel.

The two characteristics—size and weight—differ in that size can be determined from a distance, whereas differences in weight must be measured by direct contact with the nut. We must therefore be careful to observe how the squirrel actually makes choices in each case.

METHODS

In order to test for differences related to size, we will use unroasted peanuts. Sort through the nuts and find all of the nuts with only a single locule. These nuts differ from normal peanuts, which contain two seeds in the shell. Choose a site with numerous squirrels, and place pairs of nuts out for the squirrels. Be careful to place out no more pairs than can be easily watched. When a pair of peanuts is approached by a squirrel, record which nut is closest to the squirrel, which nut is picked up first, and which nut is first taken away or eaten.

The test for weight takes a little more effort. Begin by creating three classes of walnuts: light, normal, and heavy. Split each walnut carefully in half. For "light" nuts, remove nearly all of the nut from the shell, leaving a little so that there is a residual odor of nut. Glue a small piece of lead foil in each nut, and glue the nut back together. For "normal" nuts, excavate a small cavity for the piece of lead foil and glue the nut back together. Make twice as many "normal" nuts as "light" or "heavy" nuts. The "heavy" nuts will be split in the same manner as the other nuts. Excavate a ball-bearing size hole in the nut, place a round ball of lead in the hole, then glue the nut back together. The three groups of nuts now all smell like nuts, glue, and lead, but differ in weight. Present the squirrels with pairs of nuts: light and normal, or heavy and normal. Again, be careful to note which nut is approached first, whether one or both nuts are handled by the squirrel, whether the nut is eaten or stored, and which nut is left.

ANALYSES

Because nuts are always presented in pairs, we can analyze the results using a simple Chi-square test of the a priori hypothesis that squirrels choose each type of nut equally.

Suggested null hypotheses and statistical analyses include:

a. Squirrels approached each size nut with equal frequency: Chi-square test.
b. Squirrels handled each size nut with equal frequency: Chi-square test.
c. Squirrels removed each size nut with equal frequency: Chi-square test.
d. Squirrels ate each size nut with equal frequency: Chi-square test.
e. Squirrels approached each of the three weights of nuts equally: Chi-square test.
f. Squirrels handled each weight of nut with equal frequency: Chi-square test.
g. Squirrels removed each of the three weights of nuts equally: Chi-square test.
h. Squirrels ate (or at least opened) each of the three weights of nuts equally: Chi-square test.

Suggested graphical analyses include:

a. Bar graphs of the frequencies of approach, handling, removal, and consumption of each size nut.
b. Bar graphs of the frequencies of approach, handling, removal, and opening of each of the three weights of nuts.

INTERPRETATION

This set of experiments is designed to stimulate your thoughts about the way animals choose their food items. Think about the following questions when interpreting your analyses. Did you find any evidence that squirrels discriminated among nuts based on their size? Was there evidence that squirrels chose heavier nuts having the same apparent size? Was there any evidence that squirrels assessed size visually? Was there any evidence that squirrels assessed weight visually? Can you support the suggestion that squirrels actually weighed the nuts by picking them up and comparing weights? What are the consequences of these choices to individual squirrels? What selective pressures might these squirrel foraging patterns apply to nut design?

SUGGESTED REFERENCES

Janzen, D.H. 1970. Herbivores and the number of tree species in tropical forests. *Amer. Natur.* 104: 501-528.

Janzen, D.H. 1971. Seed predation by animals. *Annu. Rev. Ecol. Syst.* 2: 465-492.

Smith, C. and D. Follmer, 1972. Food preferences of squirrels. *Ecology* 53: 82-91.

Sork, Y.L. 1983. Mammalian seed dispersal of pignut hickory during three fruiting seasons. *Ecology* 64: 1049-1056.

8. *Foraging Patterns and Prey Choice by Guppies*

ABSTRACT

Foraging patterns of guppies on *Daphnia* are evaluated under lab conditions. The profitability of different sizes of prey is compared with the preferred prey of the fish.

INTRODUCTION

All animals must have food if they are to survive and reproduce. Many species actively hunt their food, and this requires time and energy that cannot be used for other important activities. This observation has produced the common argument that selection favors efficient foraging behaviors, an argument that forms the basis of "optimal foraging theory." The central tenant of this theory is that selection has optimized or maximized the efficiency of foraging as measured in some "currency." This currency is commonly thought of in terms of energy and is expressed in calories. Since efficiencies are ratios of benefits to time, foraging efficiency can be thought of as the ratio of calories obtained from food (calories in food less calories required to eat the food) to the time required to get the food. This ratio is thus maximized when the net benefits of the prey are great, or when the time required to get the prey is small, or both. We should thus expect predators to prefer those prey that yield the greatest food value per foraging time.

Many researchers have argued that optimal foraging theory is most valuable because it allows prediction of the foraging patterns expected of an animal. Others have pointed out that this prediction assumes many things about the "currency" used to measure food value, the nature of foraging costs, and the physiological, behavioral, and ecological constraints on an animal's foraging situation. Nonetheless, under simple, controlled conditions, the theory can be used to generate testable predictions about foraging. Here we will examine the prediction that predators prefer profitable prey by first determining the profitability of various prey to hungry guppies, then by observing foraging preferences of individual fish.

THE STUDY ORGANISMS

The predator: *Poecilia reticulata*

The common guppy, *Poecilia reticulata*, makes an admirable experimental animal for this study. Guppies are universally available and easily maintained indoors or wherever water temperatures range between 20-28° C. They are native to northern South America, Barbados, and Trinidad. They are omnivorous, but their liking for small invertebrates has led to their widespread introduction as a mosquito control. Like other members of the family Poeciliidae, guppies give birth to live young. The gestation period is from 4-6 weeks, maturity may occur as early as 8 weeks, and full adult size is usually reached by 6 months.

The prey: ***Daphnia*** **spp.**

Daphnia are aquatic crustaceans (Order Cladocera) that are available commercially, are easily cultured, and are naturally eaten by guppies. Of the many species, *D. magna* is most desirable, although virtually any species can be used. All species are filter feeders that live on algae. Young hatch from eggs and live about 1-2 months. Body size and pigmentation can be easily controlled by regulating food and dissolved oxygen in their environment. For our purposes, a healthy population that has been well aerated and fed regularly on blenderized cooked lima beans and spinach is best.

METHODS

Prepare four size classes of *Daphnia* by gently washing a high-density stock population through a series of four standard brass soil sieves having screens of 2, 1, 0.84, and 0.5 cm. Use only aquarium water during this process, and try not to injure the prey. The four classes should be clearly different from each other. You may increase the visibility of the prey by feeding them several drops of a solution of baking yeast that has been boiled (and cooled) with congo red indicator stain. If you cannot clearly recognize the size classes, you may feed each class a different colored ink (india ink, any other inks made of finely divided particles). This coloring is not necessary, however, and does introduce the confounding variable of prey visibility to the experiment.

Determining prey profitability

Divide the guppy population into six size classes differing in total length by about 5 mm. Place one fish of each size in a separate quart jar filled with aquarium water and allow it to calm itself for about 10 minutes. Use a pipette to add a single *Daphnia* to each jar and record the size of the prey. Use a stopwatch to record the time required for the fish to "handle" the prey (i.e., the time from the first bite to the last swallow, including any time during which the prey escapes or is spit out by the fish). Repeat this procedure until each fish has handled each prey size. If time permits, repeat the entire procedure for five guppies of each size and determine average handling times. The profitability of each prey size can be determined by weighing samples of the prey, or by using the following table of dry weights for *D. magna* provided by Werner and Hall (1974). Caloric values for *D. pulex* can be found in Richman (1958).

Size measurements for ***Daphnia magna.***

Screen category	2.00	1.00	0.84	0.50
Mean weight (μg)	371.00	108.00	37.00	18.00
Standard error	44.00	3.50	1.00	0.70
No. pans/animals per pan	5/25	3/50	5/125	5/200

(From Werner and Hall, 1974.)

Determining foraging preferences

Place 20 prey of each size in a 1-gallon fishbowl. Add a guppy to the bowl and record the sizes of each prey as it is eaten. When the fish stops foraging, remove it, replace the eaten prey, and repeat the experiment with another fish. Continue repetitions until 5 fish of each size have been used.

Find the average size of each fish size category by measuring 5 fish of each class on a wet paper towel with a ruler.

ANALYSES

We want to determine the foraging efficiency of guppies when feeding on different sizes of prey and to evaluate food preferences using this knowledge of efficiency. Most of these analyses involve graphics.

Suggested graphical analyses include:

a. Scatter plots of handling time as a function of prey size for each fish size category. These plots should indicate means for each prey size.
b. Plots of foraging efficiency (weight/time) as a function of prey size, again for each fish size category.
c. Frequency histograms of the mean frequency of consumed prey in each size category for each fish size. The modal size of consumed prey can be used as an indicator of fish preference.
d. A scatter plot of the optimal prey size (e.g., that with the greatest efficiency in "b" above) as a function of the preferred prey size (e.g., the modal value in "c" above), using each fish size as a separate data point. If the fish were foraging optimally, the points should cluster around a line with a slope of +1. (See the example below, drawn from a similar experiment done with sticklebacks and shrimp.)

INTERPRETATION

This set of experiments is designed to stimulate your thoughts about the ways that animals choose their food items. Think about the following questions when

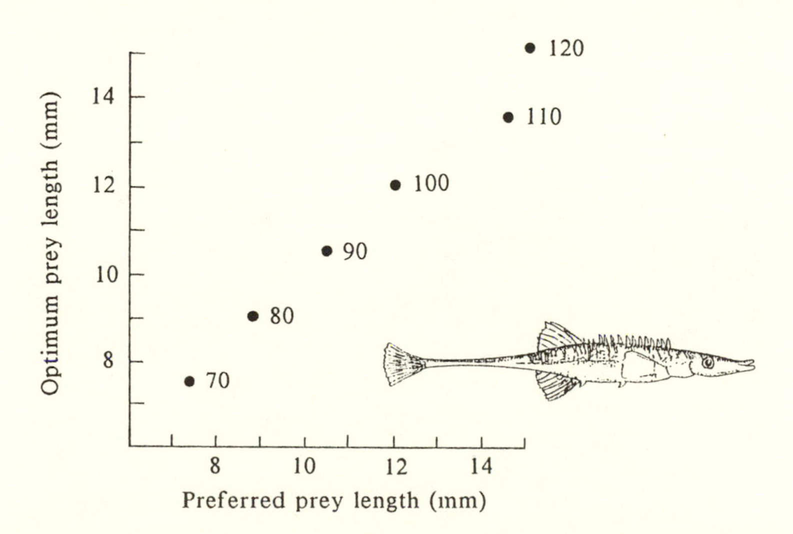

Optimum prey size (determined as dry weight of prey/handling time) as a function of the preferred prey size for 15-spined sticklebacks feeding on small shrimp in the wild. The sizes adjacent to each point indicate the size of the fish (length in mm). (From Kislalaiogu and Gibson, 1976.)

interpreting your analyses. Did larger *Daphnia* require more handling time than smaller ones? Did the guppies prefer those *Daphnia* having the shortest handling times or those yielding the greatest biomass/time? Can you rule out the possibility that the guppies chose those prey that were most obvious? What environmental factors might affect the choosiness of a guppy? What selective forces might guppy choices apply to *Daphnia* populations?

SUGGESTED REFERENCES

Kislalaliogu, M. and R.N. Gibson. 1976. Prey handling time and its importance in food selection by the 15-spined stickleback *Spinachia spinachia. J. Exp. Mar. Biol. Ecol.* 25: 151-158.

O'Brien, W.J., N.A. Slade and G.L. Vinyard. 1976. Apparent size as the determinant of prey selection by bluegill sunfish (*Lepomis macrochirus*). *Ecology* 57: 1304-1310.

Pyke, G.H. 1984. Optimal foraging theory: A critical review. *Annu. Rev. Ecol. Syst.* 15: 523-575.

Pyke, G.H., H.R. Pulliam and E.L. Charnov. 1977. Optimal foraging: A selective review of theory and tests. *Quart. Rev. Biol.* 52: 137-154.

Richman, S. 1958. The transformation of energy by *Daphnia pulex. Ecol. Manag.* 28: 273-291.

Werner, E.E. and D.J. Hall. 1974. Optimal foraging and the size selection of prey by bluegill sunfish (*Lepomis macrochirus*). *Ecology* 55: 1042-1052.

NOTES ON THE STUDY ORGANISMS

Guppies and *Daphnia* can both be purchased from biological supply houses. It is usually much cheaper, however, to obtain them from tropical fish wholesalers or pet stores. *Daphnia* are easily cultured in large jars or small aquariums. Many other species of small fish might be substituted for the predators, and brine shrimp, which can be raised from commercially supplied eggs, make an acceptable substitute prey.

9. *Seed Predation and Plant Architecture*

ABSTRACT

The foraging efficiency of seed-eating birds is investigated by manipulating the spatial patterns of seeds on artificial plants.

INTRODUCTION

Birds are both seed predators and seed dispersers. As seed predators they increase the mortality of seeds by destroying them. As dispersal agents of seeds they may increase survival of seeds by carrying them to favorable sites for germination. There is substantial evidence that plants have evolved structures that facilitate dispersal. These include structures which attach the seed to the animal (e.g. cockleburrs), fleshy coatings that attract animals (e.g. tomatoes; germination is enhanced by passage through the gut, hence the success of feral tomatoes around sewage treatment plants), and a tempo of fruit production that appears to maximize seed dispersal (e.g. Galapagos *Opuntia*). In addition, plants have evolved mechanisms that inhibit or reduce seed predation. These include protective coats (e.g. pine cones, hickory nuts), investment of the seed coat or seed with noxious chemicals (e.g. walnuts), patterns of seed distribution and production that swamp seed predators (mast years), and packaging systems that allow escape of some seeds from predators (e.g. *Croton*).

In most instances the animals that disperse seeds are also seed predators. Herbivores that feed on fleshy fruits and swallow the seeds also crush some seeds during mastication and destroy others during digestion. Thus there exists a sort of love-hate relationship between the plant and its dispersal agents, since those agents do extract a tariff for the service they provide. Many of the relationships between plants and their dispersal agents involve modification of mortality in order to insure dispersal. One attribute of this sort is the spatial distribution of seeds on individual branches. In many grasses, the seeds are found in a tight cluster or head. Birds that feed on these plants would seem to have "easy pickings." However, it may be the case that as the birds feed, they dislodge many seeds, which fall to the ground. Since it is likely that the birds can forage more effectively by moving to another seed head than by searching for the seeds that have fallen, the seeds on the ground may have escaped predation for the moment (Downhower and Racine, 1976). When seeds are solitary, predators may focus on individual seeds and attempt to eat them before moving on to another seed. Thus if both clumped and solitary seeds had the same characteristics, mortality would be higher among solitary seeds, but seed predators would be more likely to attack clumped seeds first. In this exercise we explore how the spatial distribution of seeds alters the behavior of seed predators, and how it also affects seed dispersal.

METHODS

We will examine the rates and efficiency of seed predation on two different spatial patterns of seeds. The "plants" (see figure) will consist of a dowel rod "trunk" and "branches" made from the sticks found on cotton swabs (these are wood and approximately 6 inches in length). The seeds are glued to the branches with rubber cement. Each trunk should have the same number of branches and each branch should have the same number of seeds. Half of the plants will have branches with "clumped" distributions of seeds, and half branches with "uniform" distributions of seeds. The seeds in clumps should touch one another, with all seeds attached to the branch. The seeds that are uniformly distributed should be evenly spaced along the branch. Each trunk is drilled to allow the end of the branch to be inserted. Equal numbers of plants with clumped seeds and uniform seeds are placed in a field, and a seed trap is placed around the base of each plant.

A seed trap consists of a plastic bag and three supports (see figure). The trunk passes through the bottom of the bag. The bottom of the bag is secured to the trunk with masking tape, and the mouth of the bag is held open with the three supporting stakes. Seeds and husks that are dislodged or discarded will fall into the trap. The trap will also prevent mammals from feeding on the seeds (the trap is not completely reliable in this regard, but it is fragile enough that any squirrel or mouse must tear the trap to gain access to the plant).

The "plants" should be checked daily. The number of seeds remaining on each branch are counted. These counts will be kept separately and tabulated as shown below (assuming there were 5 seeds per branch). The number of seeds in each trap is counted and recorded. Sampling should continue until more than 70% of the seeds have been removed.

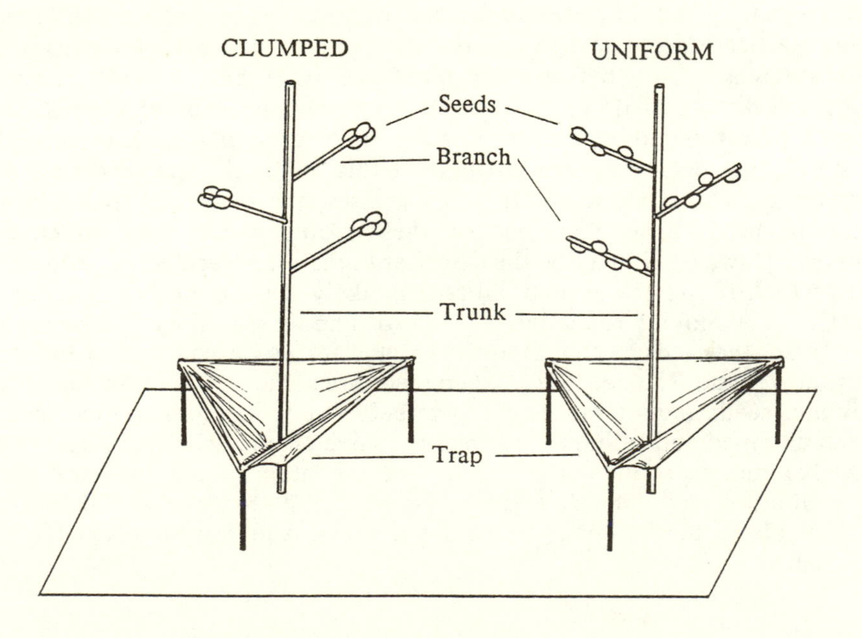

The arrangement of differing spatial patterns on seeds on "plants" constructed for this experiment.

ANALYSES

We are interested in knowing whether the patterns of seed loss and the numbers of surviving seeds are affected by the dispersal patterns of seeds on experimental plants.

Sug sted null hypotheses and statistical analyses include:

a. The number of surviving seeds remaining on the branches of the plants has a random distribution: Chi-square comparison to the Poisson distribution. (n.b., this analysis must be done separately for two types of plants.)
b. The pattern of seed survival on branches is independent of the distribution of seeds on the plant: Chi-square test; Contingency table.
c. The variability in seed survival was equivalent for each type of plant: *F* test.
d. The number of seeds dispersed and not eaten by birds was equivalent for each seed distribution: Chi-square test, a priori expectation of equivalent survival.

Suggested graphical analyses include:

a. Scatter plots of the number of surviving seeds on each plant type as a function of time.
b. Scatter plots of the number of dispersed seeds per plant type as a function of time.
c. Bar graphs of the total numbers of predated, surviving, and dispersed seeds for each plant type at the end of the experiment.

INTERPRETATION

This set of experiments is designed to stimulate your thoughts about the ways in which seed predation by birds can influence the architecture of plants. Think about the following questions when interpreting your analyses. Did the dispersal of seeds from the plant affect the number of seeds surviving attack by birds? Did birds tend to eat all of the seeds on a stem or eat seeds at random? What effect might mice or other mammalian seed predators have on the relationships you found? What other selective pressures might you imagine affecting the arrangement of seeds by a plant?

SUGGESTED REFERENCES

Cook, A.D., P.R. Atsatt and P.A. Simon. 1971. Doves and dove weed: Multiple defenses against avian predation. *BioScience* 21: 277-281.

Downhower, J.F. and C. Racine. 1976. Darwin's finches and *Croton scouleri*: An analysis of the consequences of seed predation. *Biotropica* 8: 66-70.

Goulding, M. 1980. *The Fishes and the Forest: Explorations in Amazonian Natural History*. Univ. Calif. Press.

Janzen, D. 1970. Herbivores and the number of tree species in the tropics. *Amer. Natur.* 104: 25-35.

Oppenheimer, J.R. and Lang, Q.E. 1969. Cebus monkeys: Effect on branching in *Gustavia* trees. *Science* 165: 187-188.

Racine, C. and J.F. Downhower. 1974. The reproductive biology of Galapagos *Opuntia*. *Biotropica* 6: 175-186.

10. Avian Foraging: Place and Preference

ABSTRACT

The vertical stratification and food preferences of birds are examined using a set of feeding stations suspended at different elevations.

INTRODUCTION

Competitive exclusion is the principle that each species of animal or plant must occupy its own niche. While there are many dimensions or ways in which similar niches may be separated from one another, among animals the foraging place and time and the type of food eaten are widely considered to be important axes along which niches may segregate. For example, many rather similar snails inhabit the rocky littoral zone of tropical shores. While most of these species appear to eat the same types of unicellular algae by rasping them from the rocks, they segregate clearly with respect to distance from low tide. Thus, in the Caribbean, *Nerita tesselata* is found closer to the low tide than is *N. versicolor*. Another dramatic example of this separation of species is found in the minnow communities of the streams in the midwestern United States. Several closely related species may live in the same stretch of stream. Those species that eat the same types of insects forage at different depths, while those that forage at the same depth eat different types of food.

There are a great many experiments we could conduct in an attempt to reveal niche separation. If a littoral zone (either salt or fresh water) is available we might examine the vertical zonation of snails by establishing a transect from the low to the high reaches of the zone and counting the frequency of each species in 0.5-m quadrants along the transect. If insects are active we might examine height distributions and activity times of species by hanging sticky paper (fly-paper) at different heights in the study area and examining the trapped insects throughout the day (or night). Here we use the same basic technique to examine some of the ways in which seed-eating birds avoid competition while feeding.

METHODS

A series of paired feeders will be suspended at different heights in a woodlot. The feeders, which can be made from milk cartons, will be attached to a rope that is passed over the limb of a tree (see figure). The end of the rope nearest the feeders will be tied to a stake. The rope can then be pulled taut to raise the feeders to the desired height, and the other end of the rope will be tied down to a second stake. At least five pairs of feeders should be established several days before the experiment begins to allow foraging birds to become accustomed to them. Each feeder in a pair should contain a different food item. Pairs of feeders should contain the same food items. The actual foods will depend on availability and should differ in size or texture. Millet, milo, sunflower, wheat, safflower, rape, and thistle are all commonly available bird foods. Thistle and safflower make a good combination of small and large seeds.

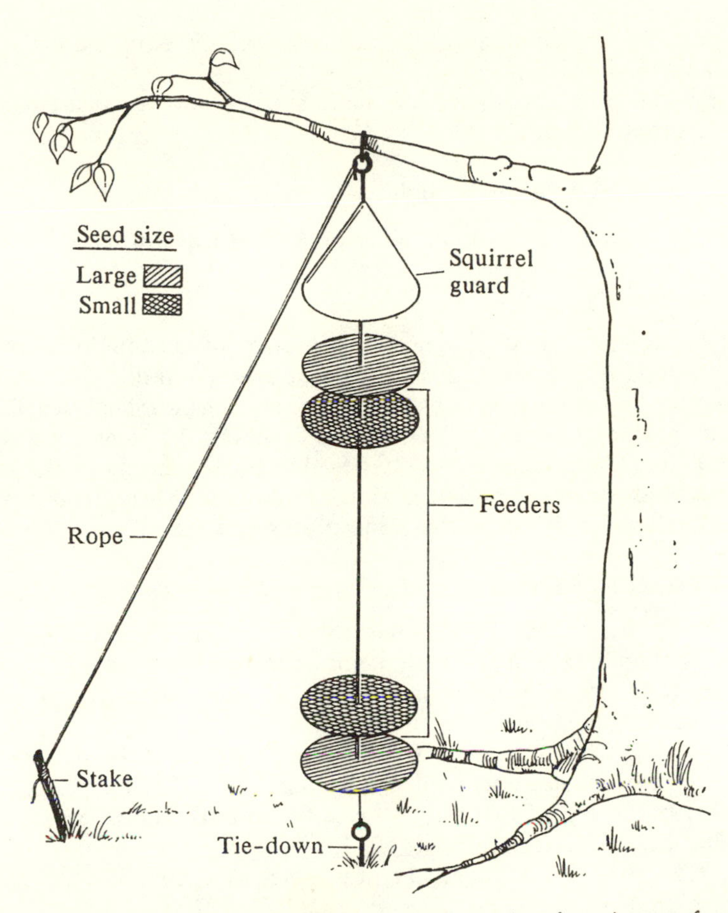

The arrangement of feeders used to examine avian foraging preferences.

Each pair of students will census the feeders using binoculars. Record the species of bird and number of individuals of each species at each feeder during 5-minute intervals separated by 10 minutes. Accumulate at least 10 censuses. This may require 2-3 days, as census intervals should be timed to associate with periods of activity of the birds, usually in the morning. You may also want to record the sex of individuals (in species that are sexually dimorphic) as well as interactions among individuals.

ANALYSES

We are interested in evaluating the preferences of different species given a choice between large and small foods at high and low elevations.

Suggested null hypotheses and statistical tests include:

a. For each species, presence or absence is independent of location or food item. This may be analyzed as a three-way table using a multidimensional contingency table (G statistic). The three variables and their states are:
   Census - present or absent;
   Location - high or low;
   Food item - large or small.
   The sample size is the number of censuses × the number of feeder stations.

b. When present, the abundance of each species was equivalent for each seed size: Mann-Whitney *U* test.
c. When present the abundance of each species was equivalent for height: Mann-Whitney *U* test.

Suggested graphical analyses include:

Bar graphs indicating the abundance of each species at each feeder.

INTERPRETATION

This set of experiments is designed to stimulate your thoughts about the ways in which animals might divide resources that are demanded by several species. Think about the following questions when interpreting your analyses. Did the birds separate along a height gradient? Did birds favoring the same seed sizes feed at different heights? Was there any evidence that birds feeding at the same height on the same seeds did so at different times of day? What conditions might allow birds to eat the same things at the same places and times?

SUGGESTED REFERENCES

Boag, P. and P. Grant. 1981. Intense natural selection in a population of Darwin's finches (Geospizinae) in the Galapagos. *Nature* 214: 82-85.

MacArthur, R.H. 1958. Population ecology of some warblers of northeastern coniferous forests. *Ecology* 39: 599-619.

Menge, B.A. 1972. Competition for food between two intertidal starfish species and its effect on body size and feeding. *Ecology* 53: 635-644.

Myers, J.P. 1980. A test of three hypotheses for latitudinal segregation of sexes in wintering birds. *Canad. J. Zool.* 59: 1527-1534.

Schoener, T.W. 1967. The ecological significance of sexual dimorphism in size in the lizard *Anolis conspersus*. *Science* 155: 474-477.

Selander, R. 1966. Sexual dimorphism and differential niche utilization in birds. *Condor* 67: 157-182.

NOTES ON THE STUDY ORGANISMS

Most locales will have enough seed-eating birds to complete this analysis at almost any time of year. The same kinds of questions asked of birds might also be asked of other animals using different techniques. For example, strips of sticky flypaper or squares of yellow plywood covered with "tangle-trap" suspended at different heights will reveal vertical stratification in insect communities.

11. Predation Rates and Search Images

ABSTRACT

The effect of a search image on predation efficiency is evaluated in an experimental system involving capture of beans by humans.

Note: Humans are the experimental animals used in this model system. If two classes can participate in this experiment, we recommend that the first set up the lab and the second provide the experimental subjects. If only one class will conduct the analyses, only those preparing the experiment should read the lab before conducting it.

INTRODUCTION

A great many studies, including several of those suggested in this text, reveal that animals rarely eat all of the prey that they encounter. Most species are selective and specialize on particular types of prey. This means that prey are rarely taken in proportion to their availability. Some studies have suggested that scarce prey are captured less than expected by chance, whereas common prey are captured more than expected by chance. A classic example of this was provided by Luuk Tinbergen, who examined the different types of caterpillars eaten by titmice during the breeding season. Tinbergen's basic finding was that caterpillar abundances changed and that each species of larva was initially rare, became more abundant during the season, and then became rare again as pupation occurred. The predatory birds appeared to avoid uncommon caterpillar species and concentrate on those that were most abundant, eating these more frequently than they should have given chance encounters. One explanation of this pattern is that the birds "knew what they were looking for." Tinbergen proposed that when a caterpillar species was rare it was uncommonly encountered by a bird. As the prey became more common, birds encountered that species more often, and learned its color and shape. The birds formed a specific **search image** of that prey, and this mental image of what the prey looked like made capture easier.

There have been other demonstrations of phenomena consistent with the formation of search images. Another classic example, provided by Reighard, involved feeding a predatory gray snapper (*Lutianus griseus*) prey fish that had been dyed blue or red. When the snapper was used to eating blue prey and was given a school of blue and red prey in equal frequency, it almost always ate all of the blue prey first. This could not be explained as a consequence of different prey palatability, and suggests that the snapper had formed an image or mental concept of what a prey looked like.

In this exercise we will examine the search image concept. While we could probably do this by feeding prey such as *Daphnia* to predators such as guppies (following the same protocol used in the snapper experiment), we can more easily do it by asking human predators to collect "prey" of different types. This we will do by presenting uninformed volunteers with beans and asking them to collect as many beans as possible.

METHODS

Obtain different types of dried beans from a grocer. The exact type of beans is unimportant, but it is necessary that they be approximately the same size and that they differ in color or pattern. Pinto beans, navy beans, and kidney beans are all commonly available and work quite well. Prepare for the experiment by sorting the beans into groups having the compositions given in the table below. One replicate of each group will be required for each experimental subject, and we would like at least 10 subjects to participate, although more are better and less will do.

Composition of "prey" populations used in the analysis of search images of human "predators." Two species of "prey" are evaluated.

POPULATION NUMBER	NUMBER OF PINTO BEANS	NUMBER OF NAVY BEANS
1	100	0
2	90	10
3	80	20
4	60	40
5	40	60
6	20	80
7	10	90
8	0	100

Store each of the groups of beans in a plastic baggie or a jar until needed. While it is possible to conduct this experiment on a table top, fewer beans will litter the floor of the lab if it is conducted in a large, shallow cardboard box. A box measuring at least 0.5 m square and several centimeters deep is fine. Whatever surface you use, make sure that it is not the same color as the beans. (For example, line the box or table top with yellow paper to highlight the beans.) Tell each experimental subject only that they are to pick up as many beans as possible during a 10-second interval. They must pick up one bean at a time and place the bean in a container (beaker or jar) next to the sampling area. Begin the experiment by dumping bean population #1 into the sampling box and spreading the beans more or less evenly over its surface. Record the 10-second interval on a stop watch. After the "predator" has "foraged" on the first population, count the number of predated beans and repeat using the second population. After each trial, count the number of beans of each type that were collected. If an adequate number of observers are available, record the order in which beans types were collected (e.g., P.P.P.N.P.P.N., etc.) for each trial.

This experiment simulates predation by a visually foraging predator on a prey population that is initially infrequent and becomes more frequent. More complicated experiments allow examination of the effects of multiple prey types. One simple modification uses the same protocol given above, but the bean populations given in the following table.

Composition of "prey" populations when three species of prey are evaluated.

POPULATION NUMBER	NUMBER OF PINTO BEANS	NUMBER OF NAVY BEANS	NUMBER OF KIDNEY BEANS
1	100	0	0
2	90	8	2
3	80	15	5
4	60	30	10
5	40	45	15
6	20	60	20
7	10	68	22

This second experiment simulates predation on one species of prey that becomes rarer and two species that are initially uncommon and become more common. Many other modifications of this same experimental design are possible and may be conducted if time permits. While you may design these yourself, one we would like to suggest involves examining the effects of cryptic coloration on foragers using a search image. This is easily done by lining your experimental arena or table top with paper that is the same color as one of the bean types (e.g., maroon to match the kidney beans). Repeat the same experiment you first performed so that you can compare the results.

ANALYSES

We are interested in knowing whether beans of each type were taken in proportion to their abundance. If they were not, we would like to know whether uncommon beans were taken less often than expected by chance and common beans were taken more often than expected by chance. The first question can be addressed statistically through application of the Chi-square test using the null hypothesis that beans were taken in proportion to abundance. This should be done by pooling the captures for each "predator" in a trial (i.e., predators combined but each trial analyzed separately). The question dealing with deviations from randomness can be addressed graphically by plotting the percentage of captures as a function of the percentage available. If the points cluster around a line with a slope of +1, then captures were proportional to availability. If the points fall below the line when beans are rare and above the line when beans are common, then foraging is consistent with formation of a search image.

INTERPRETATION

This set of experiments is designed to stimulate your thoughts about the ways that animals recognize desirable food items. Think about the following questions when interpreting your analyses. Is a search image the only way animals can increase the speed with which they identify foods? Does a search image lead to optimal foraging (maximizing the net rate of benefit from food)? What is the effect of increasing species diversity in a community on the predation rate for the most common species of prey? How might an animal learn the features of desirable prey (i.e., form the search image) in the first place? How would you explain the presence of a search image in light of the fact that unusual or "oddball" prey frequently attract special attention from predators (see Exercise 16 on fish schools)?

SUGGESTED REFERENCES

Clarke, B. 1962. Balanced polymorphism and the diversity of sympatric species. Publs. Syst. Assoc. No. 4, 47-70.

Murray, J. 1972. *Genetic Diversity and Natural Selection.* Oliver and Boyd, Edinburgh.

Popham, E.J. 1941. The variation in the color of certain species of *Arctocorisa* (Hemiptera, Corixidae) and its significance. *Proc. Zool. Soc. Lond. A* III: 135-172.

Reighard, J. 1908. An experimental study of warning coloration in coral reef fishes. Publs. Carnegie Institution No. 103: 257-325.

Tinbergen, L. 1960. The natural control of insects in pinewoods. I. Factors influencing the intensity of predation by songbirds. *Archs. Neerl. Zool.* 13: 265-336.

12. Flower Choice and Constancy in Bees

ABSTRACT

Foraging patterns of several species of bees are examined in an attempt to determine whether these species use food as they encounter it, or specialize on particular resources. Comparisons are made between species' foraging patterns to determine whether different species use different types of resources. Finally, the flower constancy of individual foragers in each species is examined in an attempt to determine the degree of foraging specialization among individuals in each species.

INTRODUCTION

A primary problem confronting animals is the location of sufficient calories and nutrients for sustenance and reproduction. Animals are frequently classified as being either "generalists" or "specialists" with regard to the types of food consumed. Generalists are thought of as taking a wide range of food items, while specialists are thought of as taking a limited range of items. Since these terms may be applied to entire species as well as to the individual organisms of a single species, common use of these terms may be confusing. For example, a generalist species may be composed of many specialist individuals.

Whether specialists or generalists, many animals feed on a range of different food sizes, shapes, colors, and species. When searching for food items, an animal may simply take food as it is encountered. In this case, the range of different food items in the diet reflects the range of available food items in the environment. Alternatively, the animal may search for particular food items and more or less ignore other potential items. In this case, the range of different food items in the diet differs from the range of food items available in the environment. The generality of this second alternative has suggested a number of hypotheses concerning animal foraging that are collected in a general "optimal foraging theory." We will examine various aspects of this theory in other labs. Here we will determine the availability of possible food resources and compare what is available with what is used by foragers of several bee species. These comparisons will also determine any differences in the foraging patterns of each species, and examine the degree to which individual foragers remain constant or faithful to particular flower types during a foraging trip.

THE STUDY ORGANISMS

In general, bees are specialized to collect pollen (and nectar) for use as a larval and adult food source. Pollen is typically gathered with the front legs and carried on small hairs on the abdomen or legs, or in specialized pollen combs on the hind legs. This means that bees carry a record of the flowers they have visited during a single foraging. Since many flowers have distinctive pollen shapes, this record can be examined by microscopic analysis of pollen grains taken from bees. On the first day of this lab, we will examine the abundance of bees on flowers in the study site, as well as the abundance of the flowers themselves. On the second day of the lab, we will examine the pollen types found on each worker bee.

FIELD METHODS

Experimenters should work in groups of 3-5, each group using a different part of the study site. Foraging bees should be collected by grasping them tightly with forceps and crushing their abdomens. Each bee should be placed in an individual vial with a couple of drops of 70% ethanol. The species of flower on which the bee was foraging should be recorded on a piece of tape wrapped completely around each vial. Since most of you will not be able to identify all flowers present in the study site, code all flowers alphabetically. For example, if the first bee is collected from *Solidago canadensis* (one of the goldenrods), it will be placed in a vial labeled "A." A specimen of the flower is then collected, placed in a plastic bag, and labeled "A." Codes will be deciphered in the lab. Collect all bees as encountered while moving through the site. Do not concentrate on any particular flower, simply grab every bee seen that is of the appropriate species (as defined by the instructor).

NOTE: IF YOU ARE ALLERGIC TO BEE STINGS, DO NOT PARTICIPATE IN THE COLLECTING PROCESS. Either record data for the collector or gather flowers for the voucher specimens.

After your group has collected about 50 bees, determine the abundance of flowers within your study area. Do this by establishing two transects 30 m long and 10 cm wide in the study area. Count all flowers that lie within the transects. Your instructor will help define a "flower," since there will be obvious differences between the types of flowers in the field. Record all flowers in the same code that your group used while collecting bees.

At the end of the lab, place all of your group's bees, together with both sets of transect data, in a large paper bag. Write the names of all group members on the bag, together with your group's number. Give the bag to the instructor, who will freeze the bees and flowers until the following lab.

LAB METHODS

Work in the same groups that collected the bees. Begin by decoding your alphabetically labeled flowers, and renaming all flower types with the appropriate Latin names (or a class code designated by the instructor). Take a stamen from each flower type, crush it on a microscope slide, and mount the pollen in a drop of water under a cover slip. Make sure that each of these sample slides is labeled. Locate individual pollen grains, and make a detailed drawing of each type, paying attention to shape, size, texture, etc. Note that many types of grains are not spherical, and may look different when viewed from different angles. After your collection of voucher slides is complete, begin examining each bee for pollen. Check pollen combs, abdominal and hind leg hairs, and the hairs of the face. Remove the pollen with a toothpick, or by washing the bee in a drop of water on a slide. Cover the drop with a cover slip, and determine the types of pollen present. For each bee, record the bee's species and the pollen types, listing the pollen types in decreasing order of abundance within the load. If no pollen is present, score the bee as having no pollen. When each bee has been examined, discard it, remove your label and wash out the vial.

ANALYSES

We are interested in three primary questions: Do the foragers in each species of bee use resources (flowers) in proportion to resource abundance, or do they tend to underuse some and overuse others? Do the different species of bees use the same resources in the same manner, or do they use slightly different types of flowers? Do the workers in each species remain constant to individual flower types during a foraging trip, and do the species differ in their degree of constancy?

Suggested null hypotheses and statistical analyses include:

a. Foragers of each species were captured on flowers in proportion to flower abundance: Chi-square test, a priori expectation.
b. Different species of bees use resources in the same proportions: Chi-square test, maximum likelihood comparison.
c. Species use of flowers is independent of flower color: Chi-square test, contingency table.
d. Mean (or median) number of pollen types collected in single foraging trips is equivalent for all species: Student's t test, Mann-Whitney U test, or analysis of variance, as appropriate.

Suggested graphical analyses include:

a. Presentation of bar graphs giving proportions of each species captured on each flower type.
b. Frequency histograms giving number (or proportion) of each species carrying different pollen types.

INTERPRETATION

This set of experiments is designed to stimulate your thoughts about the ways in which species differ in their use of food resources. Think about the following questions when interpreting your analyses: Did any of the species of bees forage randomly on the available flowers? Did all bee species show the same preferences for particular flowers, or did each species have its own preferred flower type? Were there any generalities about the type of flower that each species preferred (e.g., did one bee species prefer yellow flowers and another prefer blue)? Were there any differences in the faithfulness of the bee species to individual flower types during a foraging trip? What do these differences suggest about the willingness of the foragers to sample new resources (i.e., to run risks trying new sources of reward)? Were there differences in the total number of flower types used by bees (showing differing degrees of individual foraging constancy)?

SUGGESTED REFERENCES

Barth, F.G. 1985. *Insects and Flowers*. Princeton Univ. Press, Princeton, NJ.
Brian, A.D. 1951. The pollen collected by bumblebees. *J. Anim. Ecol.* 20: 191-194.
Brian, A.D. 1954. The foraging of bumblebees. *Bee World* 35: 61-67,81-91.
Free, J.B. 1963. The flower constancy of honeybees. *J. Anim. Ecol.* 32: 119-131.
Free, J.B. 1970. The flower constancy of bumblebees. *J. Anim. Ecol.* 39: 395-402.
Heinrich, B. 1976. The foraging specializations of individual bumblebees. *Ecol. Monogr.* 46: 105-128

Levin, D. A. and W.W. Anderson. 1970. Competition for pollinators between simultaneously flowering species. *Amer. Natur.* 104: 455-463.
Lewis, W.H., P. Vinay, and V.E. Zenger. 1983. *Airborne and Allergenic Pollen of North America.* Johns Hopkins Univ. Press, Baltimore.
Straw, R.M. 1972. A Markov model for pollinator constancy. *Amer. Natur.* 106: 597-620.
Waddington, K. and L. R. Holden. 1979. Optimal foraging on the flower selection by bees. *Amer. Natur.* 114: 179-196.
Zimmerman, M. and G.M. Pleasants. 1982. Competition among pollinators: Quantification of available resources. *Oikos* 38(3): 381-383.

NOTES ON THE STUDY ORGANISMS

Almost any bee species is a suitable subject for this study. Wasps may be abundant in the study site but may not be collecting pollen (although they are collecting nectar and insect prey) and are thus inappropriate subjects. Bee flies are mimics of bees that do not collect pollen but are commonly mistaken for bees. We do not need to identify the actual bee species, but might want to refer to a field guide to insects if many different bees are active. The procedure outlined here kills the bees humanely. Should the class want to perform this study without harming bees, we suggest trying to follow individual foragers and recording the flower species visited. Note that this modification will not separate bees collecting pollen from those drinking nectar.

13. Predation Efficiency and Prey Dispersion: Flies and Wasps

ABSTRACT

Dispersion patterns of fly pupae (*Sarcophaga*) are manipulated to evaluate effects on the efficiency of predation by *Nasonia* wasps. Prey dispersal and dispersion are examined as responses to wasp predation.

INTRODUCTION

The spatial distribution or arrangement of organisms defines a population's **dispersion pattern**. While an infinite number of arrangements are possible, it is convenient to divide possible patterns into three categories. At one extreme, the organisms might be uniformly spread over the entire environment so that each is maximally spaced from all others. This pattern is termed a **uniform** or **over-dispersed** distribution . At the other extreme, all organisms might be found in discrete clusters or clumps. This pattern is called **clumped** or **contagious**. The clumped and uniform patterns can be thought of as the endpoints of a continuum of possible distributions. If the dispersion pattern is clumped, then the presence of a single individual at a given spot increases the likelihood of that spot being occupied by other individuals. Conversely, if the dispersion pattern is uniform, then the presence of a single individual decreases the likelihood of occupation by others (as if clumped individuals attract and uniform individuals repel each other). A third dispersion pattern is generated when individuals neither attract nor repel each other. In this case, the presence of one individual has no effect on the presence of others, and the organisms are distributed randomly throughout their environment.

Many factors may interact to determine the dispersion pattern of a population. Resource distributions (e.g., feeding sites, moisture gradients, etc.) may favor a particular pattern. Certain reproductive patterns (e.g., reproduction by stolons, stem layering, budding, etc.) may determine the pattern. Alternatively, social behavior of animals may result in clumping or scattering (e.g., gregarious or territorial behaviors). The **dispersal** or movement abilities of a species are obviously interrelated with its dispersion pattern.

Dispersion patterns may reflect interactions between as well as within species. Recent attention has focused on the effects of such patterns on predation, disease transmission, and foraging ability. In this analysis, we want to examine the effects of dispersion of fleshfly (e.g., *Sarcophaga bullata*) pupae on the foraging efficiency of parasitic wasps (*Nasonia vitripennis*). We will then examine the behavior of fly larvae and the dispersion of fly pupae in the wild as possible responses to predation by parasitoids.

THE STUDY ORGANISMS

The flies we are interested in are members of the families Sarcophagidae and Calliphoridae. Most of these flies resemble houseflies; some of the Calliphorids are metallic blue or green. Larvae of most species feed on carrion, and are commonly found in road-kills and other decaying carcasses. Larvae are necessarily clumped in their dispersion, but leave the carcasses immediately prior to pupation and crawl distances of several meters. Larvae are negatively phototactic (see Exercise 1 on simple orientation movements) and eventually pupate in shaded, protected spots. Pupae are preyed upon by various vertebrate (e.g., mice) predators and invertebrate (e.g., hymenopteran) parasites which lay eggs inside the pupae. The fly *Sarcophaga bullata* is commercially available, as is the parasite wasp *Nasonia vitripennis*, suggesting that these species be used for experimentation. Other flies may also be used. (The general biology of the wasp is described in Exercise 22 on the analysis of sex ratios.)

METHODS

Dispersion and predation

Aquariums or large plastic boxes make good arenas for these analyses. Ideally, each arena should have a surface area of at least 0.25 m^2. Cover the bottom of each of three containers with a thin layer of finely shredded bark (e.g., the kind used to mulch gardens) or vermiculite. Place fly pupae on this substrate in a clumped, random, or dispersed pattern, one pattern per arena. The actual number of pupae used will depend on arena size, but must be equal for each of the three treatments. A typical 10-gallon (40-l) aquarium should hold six pupae. For the clumped treatment, use a random number table to choose one corner of the tank and place all six pupae in the corner. For the uniform treatment, place a pupa in each corner and one midway along each side of the tank. For the random treatment, use the dimensions of the tank to form an "X" and a "Y" axis. Use a random number table to obtain six sets of Cartesian coordinates, and place a pupa at each of these points. Replicate each of the three treatments at least five times. Place a single fecund female wasp in each arena and cover with a sheet of glass. After 24 hours, remove all pupae, and place in a labeled vial to incubate at room temperature. (Make sure you do not include the wasp in the vial.) Eight to thirteen days later, dissect the pupae and record the number that contained flies or wasps. Alternatively, allow the insects to enclose and record the number of pupal cases that split open (flies emerged) and the number that have small exit holes (wasps emerged).

Fly behavior, dispersal, and dispersion

Analyses of the dispersal of fly larvae are fairly straightforward and can be conducted in the laboratory. Analyses of dispersion patterns must be conducted outdoors, in a study area free from disturbance by humans and scavengers.

Dispersal. Fill a 30-foot-long segment of vinyl or aluminum guttering with coarse, moist sawdust or fine vermiculite to a depth of about 3-4 cm and apply a very thin film of petroleum jelly to the inside upper edge of the sides and ends. Prepare a colony of flies, either by allowing oviposition on a ball of meat (1/2 kg of kidneys, liver, etc.) or by inoculating a ball of meat with several hundred growing larvae. Place the colony at one end of the trough and check it daily until it is apparent that larvae are dispersing. Inoculate the trough with parasites by placing 20 parasitized pupae next to the colony. Mark each parasitized pupa with a dot of white typewriter correction fluid so that it can be distinguished from dispersing larvae. Do not allow the sawdust to dry completely. After dispersal is complete and pupation has occurred, remove the sawdust from 0.1-m sections of the trough, recording the distance from point of origin and number of pupae per section. If time permits, weigh the pupae in each section to obtain an average weight as an estimate of body size. Dissect the pupae to determine whether parasitism has occurred, or place each decimeter's pupae in a container and allow the wasps to complete their life cycle. Pupae may be easily sorted from sawdust either by hand or by passing the mix through a soil sieve fine enough to retain larvae and coarse enough to pass sawdust.

Dispersion. The study of dispersion patterns of larvae in the field is complicated by effects of weather, local substrate variation, and scavenging by dogs and wild animals. The simplest experimental design involves enclosing a colony of fly larvae in a 1/2-inch mesh hardware cloth box. Anchor the box in the center of a grassy area using large spikes or iron rod (e.g., rebar). Arrange 100 square tiles (e.g., linoleum floor tiles) in a circle 8 m in diameter with the colony at its center (i.e., tiles are placed about 0.25 m apart). The exact size of the tiles will depend on availability, but each should measure at least 20 × 20 cm. Allow the colony to disperse, checking the tiles daily until pupae are obvious beneath the tiles. Count the numbers of pupae beneath each of the tile quadrats.

ANALYSES

In the first part of this analysis we are interested in the interaction of prey dispersion with predator efficiency. In the second part we are interested in prey dispersal and dispersion, and the effectiveness of predation on dispersed prey. (For a discussion of the analysis and interpretation of insect dispersal, see exercise 15 on dispersal and dispersion in soldier beetles.)

Suggested null hypotheses and statistical analyses include:

a. The fate of pupae was independent of pupal dispersion: Contingency table.
b. The frequency distribution of dispersal distances was half-normal: Analysis of kurtosis.
c. Wasps parasitized dispersed pupae in proportion to pupal abundance: Chi-square test, a priori expectation of equal proportions.
d. Larvae dispersed themselves randomly beneath sample quadrats: Goodness of fit comparison of observed and Poisson distributions.

Suggested graphical analyses include:

a. Frequency histograms of dispersal distances and predation success for pupae removed from the trough.
b. Comparison of the frequency distribution of pupae per quadrat with that expected by a Poisson distribution.

INTERPRETATION

This set of experiments is designed to stimulate your thoughts about the consequences of predation to the spatial arrangement of animals. Think about the following questions when interpreting your analyses. What affect did the dispersion pattern of the prey have on predation rate? Did all of the larval flies show the same propensity to disperse? Could you detect any effect of larva size on dispersal distance? Did larvae that dispersed farther avoid predation more often? Did you find evidence that the larvae avoid one another when pupating? What effect might you expect other types of predators (e.g., shrews or mice) to have on dispersal and dispersion?

SUGGESTED REFERENCES

Crook, J.H. 1966. The adaptive significance of avian social organization. *Symp. Zool. Soc. Lond.* 14: 181-218.

Croze, H.J. 1970. Searching image in carrion crows. *Zeits. Tierpsychol.* 5: 86.

Horn, H.S. 1968. The adaptive significance of colonial nesting in the Brewer's blackbird *Euphagus cyanocephalus. Ecology* 49: 682-694.

Lack, D. 1954. *The Natural Regulation of Animal Numbers.* Clarendon Press, Oxford.

Tinbergen, N., M. Impekoven and D. Franck. 1967. An experiment on spacing out as a defense against predators. *Behaviour* 28: 307-321.

NOTES ON THE STUDY ORGANISMS

Both the fly larvae and the wasps can be obtained from biological supply houses. Houseflies may be used as prey and are easily grown. A shrew may be substituted for the wasps to examine the effects of a vertebrate predator.

14. *Behavioral Aspects of Resource Partitioning*

ABSTRACT

Small bumblebee workers are observed foraging on goldenrod inflorescences when solitary and when in the presence of large bumblebee workers. Foraging patterns are quantified and changes in behavior attributable to competitive interactions are analyzed.

INTRODUCTION

One important ecological principle is that of **competitive exclusion**. This principle states that only one species may fill a single ecological niche for a prolonged period. If two species have overlapping niches, as when they use some common resource, they will be in competition. This principle has frequently been used to explain the observation that two or more species appearing to use the same resource generally partition the resource by using different parts of it. For example, several species of closely related minnows may inhabit the same length of a stream but use different depths. Similarly, minnows using the same depths in a stream may eat different foods.

Behavioral interactions between individuals have commonly been thought to affect the degree of resource partitioning. For example, a species may be constrained in its use of a resource when competitors are present, but have access to the resource when competitors are absent. In this lab we will examine behavioral interactions between two groups of bumblebees foraging on goldenrod (*Solidago* spp.), a common and abundant source of pollen and nectar in late summer and early autumn. Specifically, we will examine the foraging patterns of small bumblebee workers when competitors are present, and when competitors are absent. Our goal will be to determine whether behavioral interactions between these groups change the foraging patterns of the small bees.

METHODS

We will duplicate the experiments of Morse (1977). Choose several flowering stems of goldenrod, making sure to use a species with elongated inflorescences (e.g., *Solidago canadensis*). Select inflorescences that have as many of their flowers open as possible (preferably all of them). Each inflorescence can be thought of as being composed of three sections: proximal (adjacent to the stem), medial (the middle), and distal (furthest from the stem). (This relationship is shown diagrammatically in the accompanying figure.)

Place a screened cage over the inflorescences. The cage should be made of wire screening and be approximately 1 m in diameter and 1 m tall. Capture a small bumblebee worker that is foraging on goldenrod, and release the bee in the cage. When the bee starts foraging, score the location of the first 50 florets (individual flowers in the inflorescence) that the bee visits. Do this numerically by assigning a score of 3 for each proximal, 2 for each medial and 1 for each distal floret. After scoring 50 visits, release 3-5 large bumblebee workers in the

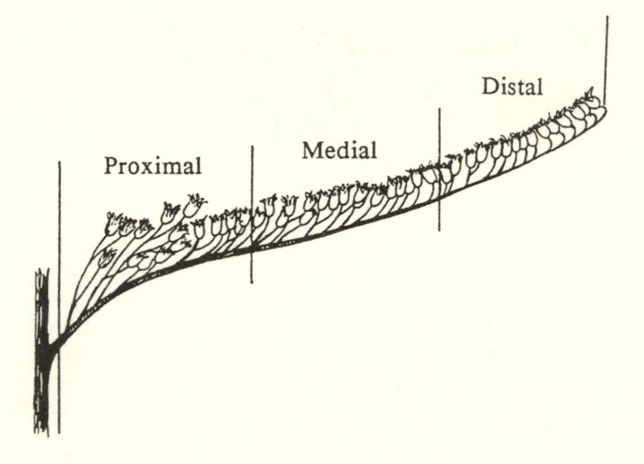

Solidago canadensis *inflorescence divided into proximal, medial and distal segments.*

cage. Wait until the small bee and one of the large bees come within 20 cm of each other, and then score the location of the next 50 florets visited by the small bee. When both sets of scores have been completed, release the bees and begin again with a newly captured small worker. Continue this routine until the class has at least 10 small worker observations with and 10 without large bees.

ANALYSES

We are primarily interested in comparing the average floret visited by small workers when competitors are present with that when competitors are absent.

Suggested null hypotheses and statistical analyses include:

Use a Mann-Whitney *U* test to compare medians or a student's *t* test to compare means. Treat each bee as a single observation by summing the total score for its 50 florets, yielding a total score greater than or equal to 50 and less than or equal to 150 for each bee.

Suggested graphical analyses include:

Bar graphs of either the number of visits or the percent of visits to each floret position.

INTERPRETATION

This set of experiments is designed to stimulate your thoughts about the effects of competition between species on the foraging patterns and social interactions of the species. Think about the following questions when you interpret your analyses: Were the foraging patterns of each species the same or different when no competitors were present? What was the effect of presence of a socially dominant competitor on the socially subordinate species? What was the effect of presence of a socially subordinate competitor on the socially dominant species? Why would a species change its foraging pattern when a competitor is present (i.e., what's the payoff, since it loses a resource)? What are the consequences of these interactions to community structure? How do the effects of competitors compare to the effects of predators (i.e., resource depression; see Stein and Magnuson, 1976)?

SUGGESTED REFERENCES

Brian, A.D. 1957. Differences in the flowers visited by four species of bumblebees and their causes. *J. Anim. Ecol.* 26: 41-78.

Grant, P.R. Experimental studies of competitive interactions in a two species system: The behavior of *Microtus*, *Peromyscus*, and *Cleirionomys* species. *Anim. Behav.* 18: 411.

Heinrich, B. 1976. Resource partitioning among some eusocial insects: Bumblebees. *Ecology* 57: 874-889.

Johnson, L.K. and S.P. Hubbell. 1975. Contrasting foraging strategies and coexistence of two bee species on a single resource. *Ecology* 56: 1398.

Morse, D.H. 1974. Niche breadth as a function of social dominance. *Amer. Natur.* 108: 818.

Morse, D. 1977. Resource partitioning in bumblebees: The role of behavioral factors. *Science* 197: 678-680.

Stein, R.A. and J.J. Magnuson. 1976. Behavioral responses of crayfish to a fish predator. *Ecology* 57: 751-761.

15. Dispersal and Dispersion in Soldier Beetles

ABSTRACT

Mark-recapture methods are used to examine movement patterns in a population of soldier beetles (*Chauliognathus pennsylvanicus*). Dispersion patterns of the beetles and their principal food sources are examined through quadrat sampling.

INTRODUCTION

The dispersal movements and spatial arrangement of individuals may be important determinants of population density, persistence, and gene flow within and between groups. Dispersal and dispersion are frequently consequences of behavioral interactions between individuals; thus analysis of the patterns themselves and of the behavioral interactions of individuals are important to complete understanding of the behavioral ecology of an organism.

There are several approaches to the analysis of dispersal movements. A common method involves summarizing dispersal distance measurements in a frequency histogram and finding some simple equation that describes the shape of this curve. Many different equations have been found to adequately describe the movements of various species; one family of equations that has been used extensively has the form

$$\text{Frequency} = a + b\,(\text{Distance})^c$$

When the exponent c is 2, the curve is said to be **half-normal**, that is, it represents one-half of a bell shaped curve. This type of curve might be generated if all individuals moved in equal dispersal intervals and directions were random (for example, fleas that hop equal distances, and have equal probabilities of hopping either forward or backward with each jump). This type of dispersal is analogous to simple gas diffusion. As the exponent becomes smaller, the distribution becomes more **leptokurtic**; that is, more individuals moved shorter distances and more moved longer distances than expected for a normal curve. This increase in kurtosis (see figure) implies that organisms differ in their dispersal movements, either because of differences between the animals themselves or because of differences in distance to acceptable stopping places in the environment.

Several methods have been used to evaluate the exponent c and solve for the constants a and b. Nonlinear regression has been used to solve for the constants. Alternatively, various values of c have been plugged into linear regression equations (e.g., $c = 2, 1, 1/2...$). Finally, the kurtosis of the frequency histogram has been evaluated directly by calculating the appropriate statistics for kurtosis. The general result of all these approaches is demonstration of some degree of leptokurtosis. Normal distribution of dispersal distances is apparently exceptional in nature.

Dispersion patterns can also be analyzed in many ways, some of which are detailed in the statistics section of this manual. The goal of these alternatives is to identify the degree of clumping of organisms, and ultimately to demonstrate

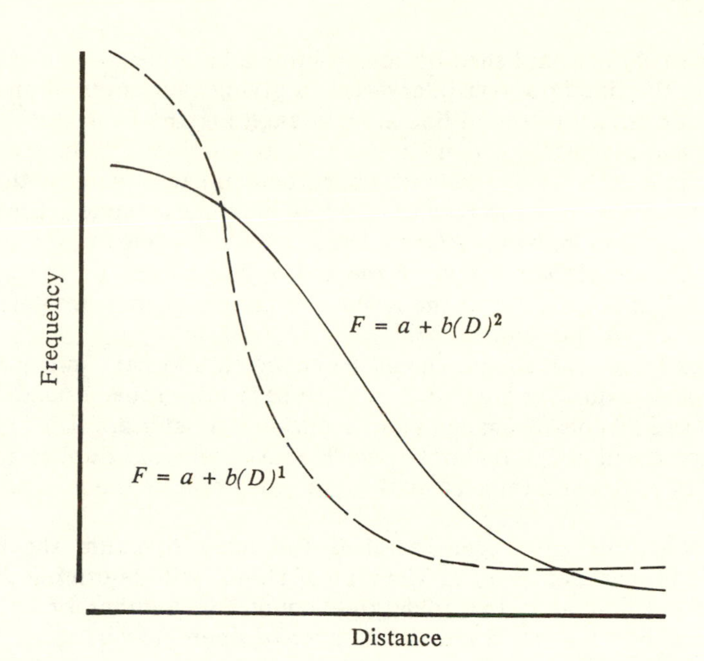

Examples of a half-normal (solid line) and a more leptokurtic (dashed line) frequency distribution.

the reasons and consequences of spatial patterning.

In this lab we will examine movement patterns in a population of beetles. While any number of beetles lend themselves to this type of analysis (e.g., Japanese beetles, blister beetles, coppersmith beetles), we will use soldier beetles, *Chauliognathus pennsylvanicus*, a species whose general biology is described in Exercise 2 on assortative mating. Specifically, we will characterize the shape of the dispersal histogram, look for differences among males and females, and quantify the dispersion patterns of the beetles and their primary food plant. Advanced students may want to use partial correlation analyses to examine correlates of beetle abundance and food plant density.

METHODS

This project is necessarily divided into two discrete parts: marking and recapturing the beetles. The two parts will yield most data if they are separated by about 24 hours. A separation of several days is acceptable, although recaptures will be much less frequent.

Marking the beetles

The study site should have a high density of beetles and food plants. Depending on the time of the year, suitable plants may include *Bidens* spp. and goldenrod (*Solidago* spp.), and analyses will be simplified if the study site has little or no flower diversity.

Divide the study site into a grid composed of 3 × 3 m squares or **quadrats**.

This is most easily accomplished by constructing a baseline along one side of the site, dividing the line into 3-m intervals, and giving each interval an alphabetic designation. Construct a second line at right angles to the first, divide it into 3-m intervals, and give each of them a numeric designation. Use tape measures or strings to complete the grid, and mark the corners of each square with wire stake flags. Each square will be referred to by its alphanumeric name representing the row-column position in the grid. Once that is done, the name of each quadrat will be defined by the alphanumeric of the lower left corner of the square. For example, quadrat A1 will be in the lower left corner of the complete grid. The figure below shows this arrangement.

The beetles from each square should be collected and held in a jar. When all the beetles in that quadrat are captured, give each individual a quadrat-specific mark composed of dots of enamel paint 1 mm in diameter applied to the elytra. Your instructor will assign quadrat-specific marks to avoid duplication. Record the number of males and females marked and then release them in the center of the quadrat.

After all quadrats have been searched and marked, count the number of available food plants per quadrat. The type of flower will determine the methods used to make this count; for goldenrod, count the number of inflorescence bearing stems. For *Bidens*, count the number of open flowers.

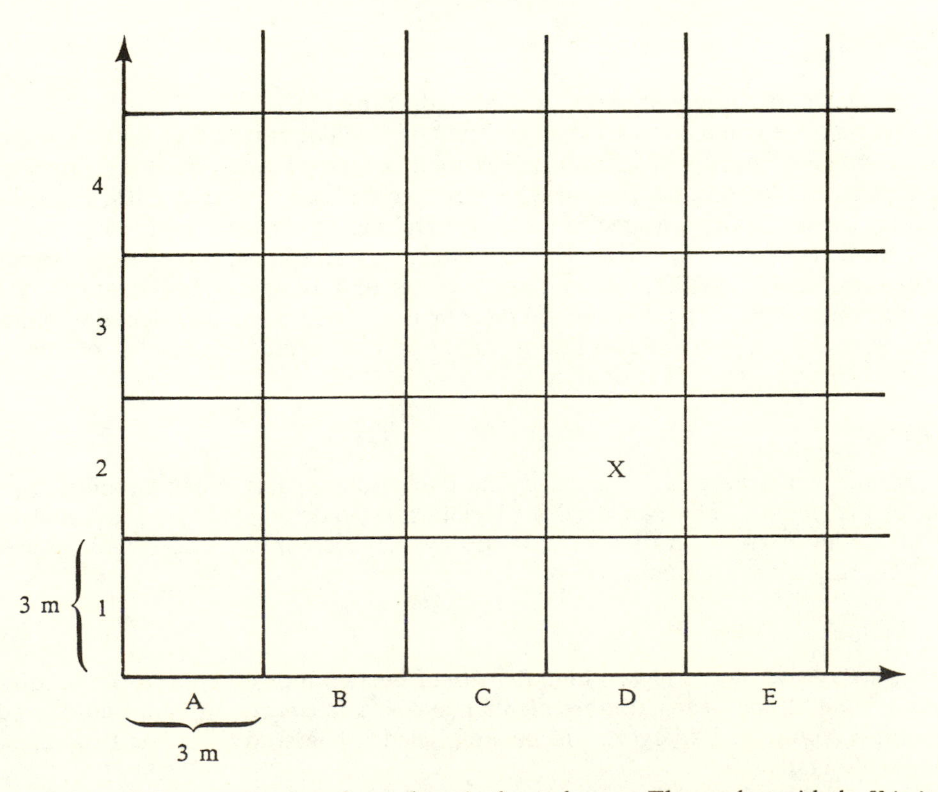

Map of the grid demarcating sample quadrats in the study area. The quadrat with the X in it is named D2.

Recapturing the beetles

Collect all the beetles from each quadrat and place them in a suitable jar that has been labelled with the quadrat's name. Return them to the lab and either anaesthetize them or freeze them. Record the mark and sex for each marked beetle, and the number of beetles of each sex that are unmarked.

ANALYSES

Dispersal analyses will use the distances moved by each individual. These distances can be most conveniently calculated by assuming that beetles moved from the center of the quadrat in which they were marked to the center of the one in which they were recaptured. Obviously, we do not know what path the beetles actually took, but this assumption allows us to work with minimum distances, and to calculate distances using the Pythagorean theorem. Means and variances for histograms should be constructed using convenient class intervals (e.g.,10-m classes). The two sexes can be compared and, if not different from each other, combined for further analysis. Kurtosis can be calculated directly. Flowers per quadrat can be evaluated using a goodness of fit test and the Poisson distribution.

Suggested null hypotheses and statistical analyses include:

a. Median dispersal distances for each sex were equivalent: Mann-Whitney *U* test.
b. Distance was not a function of beetle size: Linear regression.
c. Shapes of the frequency distribution for each sex were equivalent: Chi-square test, maximum-likelihood comparison.
d. Frequency distributions of movements were leptokurtic: Skewness and kurtosis.
e. Beetles (flowers) were randomly distributed in the field: Chi-square test, goodness of fit comparison to the Poisson distribution
f. Beetle densities were a function of flower densities: Linear regression, correlation, partial correlation.
g. Densities of beetles of one sex were a function of the densities of beetles of the other sex: Linear regression, correlation, partial correlation.

Suggested graphical analysis include:

a. Frequency histograms of dispersal distances for each sex.
b. Bar graphs of the number of beetles (flowers) per quadrat and those expected given random dispersion.

INTERPRETATION

This set of experiments is designed to stimulate your thoughts about animal movement patterns. Think about the following questions when interpreting your analyses: Did both sexes show the same dispersion patterns? Was there any effect of body size on dispersal distance? Were the two sexes recaptured at the same frequency; if not, what does this tell you about the activities of the sexes? Was there evidence that all beetles made the same kinds of dispersal movements? Was there evidence that the beetles were looking for particular kinds of stopping points (e.g., flowers in full bloom, or flowers with lots of members of the opposite sex visiting them)? What can you conclude about the rate at which genes might be exchanged between soldier beetle populations? What can you conclude about the size of the area occupied by a soldier beetle population?

SUGGESTED REFERENCES

Brown, L. and J. Brown. 1984. Dispersal and dispersion in a population of soldier beetles *Chauliognathus pennsylvanicus*. *Environ. Entomol.* 13: 175-178.

Dobzhansky, Th. and S. Wright. 1943. Genetics of natural populations. X. Dispersal rates in *Drosophila pseudoobscura*. *Genetics* 28: 304-340.

Freeman, G.H. 1977. A model relating numbers of dispersing insects to distance and time. *J. Appl. Ecol.* 14: 477-487.

McCauley, D.E., J.R. Ott, A. Stine and S. McGrath. 1981. Limited dispersal and its effects on population structure in the milkweed beetle *Tetraopes tetraophthalmus*. *Oecologia* 51: 145-150.

Poole, R.W. 1947. *An Introduction to Quantitative Ecology*. McGraw-Hill, New York.

Taylor, R.A.J. 1978. The relationship between density and distance of dispersing insects. *Ecol. Entom.* 3: 63-70.

NOTES ON THE STUDY ORGANISMS

While this analysis suggests the use of soldier beetles, many other small animals can be examined. Many species of beetles or bugs are locally abundant and make good experimental organisms if they can be marked for recapture. Freshwater or littoral zone marine snails make admirable experimental animals for this analysis. Animals ranging from isopods to small benthic fishes (like the sculpin, *Cottus bairdi*) may all be examined using the same basic procedures.

16. Fish Schools: Social Facilitation of Feeding and Predator Avoidance

ABSTRACT

The hypothesis that schooling behavior facilitates location of food is tested by allowing fish schools of various sizes to search for food hidden in a maze. The hypothesis that schools confuse predators is tested by allowing predation on schools of various sizes and composition.

INTRODUCTION

The structure and function of fish schools have received attention from both evolutionary biologists and fisheries managers. Interest from behaviorists reflects the facts that schools show extreme degrees of social interaction and organization. Interest from fisheries biologists reflects the fact that most commercially important marine fishes school. Diversity in the interest of those studying schooling has resulted in diverse approaches to the phenomenon, and even the definition of "schooling" has remained controversial. Most workers agree on these general points: A school is a temporary group of individuals; it is typically composed of fish of uniform size, shape, color and age; it is maintained by mutual interactions among the individuals composing it; and it results in the display of organized actions by its members.

Many benefits have been attributed to schooling. Two of the principal potential benefits involve protection from predators and improved feeding ability. Predators may be more easily detected by a school, and individual members may reduce their risks of predation, either because a large school confuses a predator (confusion hypothesis), because the school resembles a large individual fish (superorganism hypothesis), or because conspecifics make good hiding places (selfish herd hypothesis). Food may be more easily located because many sense organs can search a broad area in a short time. At the same time, hydrodynamic considerations may allow schooling fish to have lower metabolic costs than single individuals. Finally, schooling means that conspecifics will always be nearby; thus reproduction can occur without lengthy searches for mates.

Here we will examine two potential benefits of fish schooling. In the first set of experiments we evaluate the effects of school composition on prey location and feeding. In the second set of experiments we evaluate the hypotheses that schools decrease predation through a confusion effect, and that this decrease favors schools of fish that all look alike.

METHODS

Effects of school size on feeding

Experiments will be conducted in plastic wading pools with diameters of approximately 2 m. Before filling the pools with water, arrange 100 opaque plastic cups (e.g., condiment cups from fast-food restaurants) in a circle so that they are approximately 10 cm apart and 10 cm from the wall of the pool. Glue the cups to the pool bottom with silicone adhesive. Fill each pool to a depth of about 10 cm, and place an open-ended cylinder (e.g., a coffee can with both ends removed) in the pool's center. The cylinder must extend above the water line. Randomly pick one cup in the experimental array and place a small amount of fish food (e.g., chopped earthworms or frozen tubifex worms) in it. Place the fish to be tested in the central cylinder. Make sure that all fish have fasted for 48 hours prior to the test. After they have recovered from their transfer (about 1 minute), raise the cylinder and release the fish. Record the time from release to feeding for the first fish to feed and for a "focal point" fish that you selected prior to release. Test several replicates of each of the following school sizes: 1, 2, 4, 6, 10, 15, 20, 30. (Note that the "school" of size 1 has only one fish and is not really a school.) Larger schools will be harder to follow unless the focal point fish is distinctive. The easiest way to follow single individuals is to use a species of test fish that has color morphs. While any small schooling or shoaling species could be used, juvenile goldfish, *Carassius auratus*, are easily obtained, tolerate lab temperatures and handling, and school well. Furthermore, it is easy to compose a school of distinctly colored individuals, allowing recognition of the focal point fish.

Effects of school size on predation

Experiments can be conducted in the same wading pools used previously. Pools should be filled to approximately 20 cm. Predators should be placed in each pool at least 24 hours prior to experimentation. While many species might be used as predators, large mouth bass (*Micropterus salmoides*) will eat small goldfish and perform well at room temperatures. Regardless of the species used, the predator must be acclimated to the tank and habituated to humans. If it is terrorized during the study it will refuse to eat. Prey schools should be released from the central cylinder as described above. The stock population of prey should be composed of both orange and melanistic fish so that the following schools can be exposed to the predator: solitary fish, 4 fish of one color together with a fish of the other color, 9 fish of one color together with a fish of the other, 19 fish of one color together with a fish of the other. If the predator will eat the prey, record the time to first capture and the first prey's color, remove all other prey and introduce the next school. If the predator refuses to eat immediately, check periodically to find the first prey eaten.

ANALYSES

We are primarily interested in two questions: Does school size influence the speed with which individual fish locate prey, and does school size influence the safety of school members? Regarding the first question, we might expect that larger schools can search an area for food more quickly, and that school members will be attracted by the sight of any foraging conspecific. Regarding the second question, we might expect that predators have greater capture times when confronting larger schools (due to confusion), but that any outstanding individual might reduce predator confusion and thus fall easy prey.

Suggested null hypotheses and statistical tests include:

a. Mean times to food location by the first forager in each school were equivalent for schools of different sizes: Analysis of variance.
b. Mean times to food location by the focal fish in each school were equivalent: Analysis of variance.

c. Mean times to prey capture by a predator were equal for all schools: Analysis of variance.
d. Frequency of predation of prey that looked like the rest of the school was equivalent to that for prey that looked different from the rest of the school: Chi-square test, a priori expectation that all prey have equal risk.

Suggested graphical analyses include:

a. Scatter plots of the times taken for the first fish and for the focal point fish to locate the food.
b. Bar graphs showing the observed frequency of "oddball" prey taken from schools of different sizes. These are to be compared to bar graphs showing the expected frequency of prey taken from schools of different sizes. A sample bar graph from a study of a different species is illustrated.

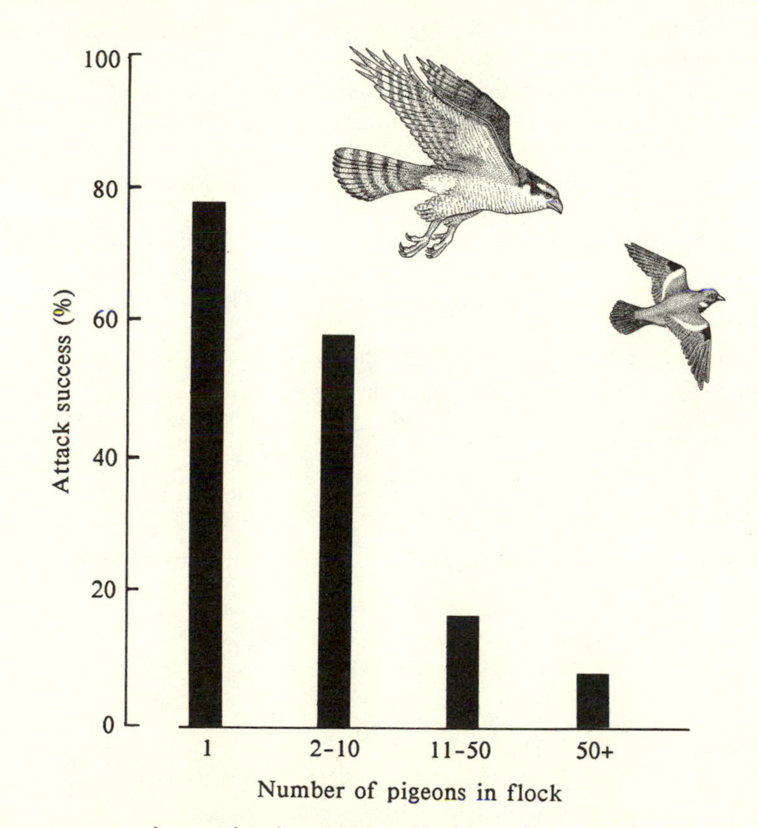

The success rate of a goshawk attacking flocks of woodpigeons of various sizes. Successes were much rarer in large flocks, probably because the goshawk could not single out an individual prey bird.

INTERPRETATION

This set of experiments is designed to stimulate your thoughts about the benefits of living in a group. Think about the following questions when interpreting your analyses: Was there evidence that individual fish in larger schools found food more quickly? Was there evidence that members of a school were attracted to other members who were feeding? Were predators more successful when attacking larger schools? Were predators more successful when they could single out an "oddball" fish? Why are fish schools characteristically composed of fish of the same species, size, and color? Why are schools of fish on coral reefs characteristically composed of fish of the same size and color but not species? What do your findings tell you about the hypothesis that predators are "prudent" and eat the prey that have the lowest reproductive value, thereby insuring a continued prey population?

SUGGESTED REFERENCES

Breeder, C.M. 1967. On the survival value of fish schools. *Zoologica* 52: 25-40.

Cushing, D.H. and F.R. Harden-Jones. 1968. Why do fish school? *Nature* 218: 918-920.

Keenyleside, M.H.A. 1955. Some aspects of schooling behavior in fish. *Behaviour* 8: 183-248.

Landeau, L. and J. Terborgh. 1986. Oddity and the "confusion effect" in predation. *Anim. Behav.* 34: 1372-1380.

Morrow. 1948. Schooling behavior in fishes. *Quart. Rev. Biol.* 23: 27-38.

Ohguchi, I. 1981. Prey density and selection against oddity by three-spined sticklebacks. *Adv. in Ethol.* 23: 1-80.

Partridge, B.L. 1980. The effect of school size on the structure and dynamics of minnow schools. *Anim. Behav.* 28: 68-78.

Partridge, B.L. 1980. *Structure and Function of Fish Schools.* Oxford Univ. Press, Oxford.

Partridge, B.L. 1982. The structure and function of fish schools. *Sci. Amer.* 246(6): 114-123.

Pitcher, T.J. and A.E. Magurran. 1983. Shoal size, patch profitability and information exchange in foraging goldfish. *Anim. Behav.* 31: 546-555.

Shaw, E. 1962. The schooling of fishes. *Sci. Amer.* 206(6): 128-138.

NOTES ON THE STUDY ORGANISMS

Predation can be examined using colored *Daphnia* instead of fish. (See Ohguchi, 1981.)

17. *Sexual Differences in Larval Construction and Placement in the Evergreen Bagworm*

ABSTRACT

Differences in larval construction and pupation sites of male and female bagworms are quantified. These differences reflect differential habitat use, and apparently influence the reproductive success of each sex.

INTRODUCTION

Many animals occur as either males or females. Obviously each type needs the other when reproducing, hence the sexes (or at least their gametes) must encounter each other when reproducing. At other times, however, the sexes may share the same habitat, or they may live quite separately. Females of many temperate bat species, for example, raise their young in maternity colonies, while males spend their summers elsewhere. Female rubythroated hummingbirds spend their summers at higher elevations than do males. Female elephant seals spend their winters at lower latitudes than do males. Whatever the behavioral, morphological, or physiological causes of these differences, one consequence of such separation is reduced competition between males and females for resources like food and shelter.

Although sexual differences in habitat preferences reduce competition and increase available resources, they are not universal. The benefits associated with group living may favor habitat sharing by the sexes; the necessities of biparental care may require habitat sharing; and constraints on mate location resulting from short lifespans and limited mobility may select for close proximity of the sexes. In such situations males and females can still avoid competition with their mates by subdividing their shared resources. Many examples of sexual resource partitioning are known. Male *Anolis* lizards, for example, choose larger diameter perches that are more elevated than do females of the same species. Male red-eyed vireos forage among higher branches and use different prey capturing techniques than do their mates. Male peregrine falcons forage in different areas and capture different sizes and species of prey than do their mates. Finally, many dramatic cases of sexual dimorphism, especially in beak sizes of birds, are attributable to sex-dependent food habits (see figure on page 66).

THE STUDY ORGANISMS

We could examine sexual differences in the behavior of practically any species by measuring such things as food particle size, food type (species), movement, perch size, perch position, home range, or almost any other quantifiable aspect of behavior. The evergreen bagworm (or one of the other common bagworms) makes a good study organism because it is abundant, easily studied, and shows clear sexual differences in cocoon construction and pupation site that may be expected to affect the fitness of males and females. Here we examine such differences.

The evergreen bagworm, *Thyridopteryx ephemeraeformis*, is a pyralid moth found throughout eastern North America; it is being replaced by similar species

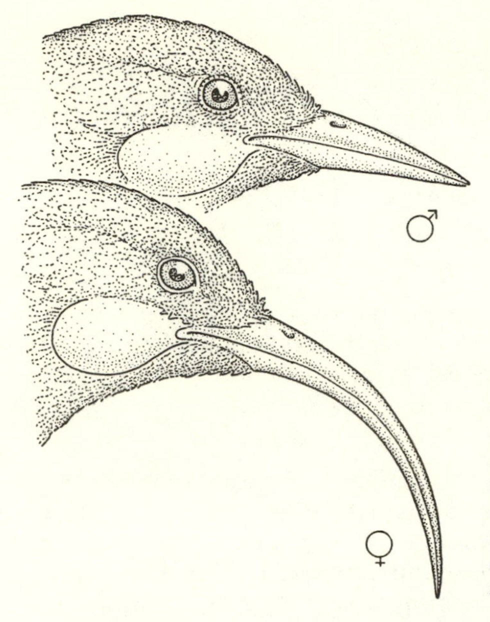

Dramatic sexual dimorphism in beak shape in the New Zealand huia. These differences presumably reflect different selective pressures due to sex-specific feeding patterns. The male's bill appears to be adapted to excavate for insects beneath bark, while the female's bill appears adapted to probe deep into insect tunnels. (From Alcock, Animal Behavior, *1975.)*

over the rest of the continent. Diapausing eggs overwinter within their mother's cocoon, suspended from a twig of a deciduous or evergreen tree. Larvae emerge in spring and "balloon" to new sites on fine threads of silk. Each larva spins a silk cover around its body, protrudes its head and legs from the anterior of this bag, and feeds on the foliage of its host tree. As the larva grows it enlarges its bag, and may decorate it with bits of leaf or cut petioles and stems.

In autumn, the caterpillars permanently suspend their bags from twigs on the host and pupate. Males emerge as strong-flying but short-lived adults. Females metamorphose into wingless, wormlike reproductives that remain in their cocoons. Males locate mates by following pheromone plumes from the females, and mating occurs in the female's cocoon. After laying her eggs in the cocoon, the female dies.

Despite its common name, the evergreen bagworm is an extreme generalist which can be found on virtually any deciduous or evergreen tree or shrub. The host species determines the obviousness of sexual dimorphism, at least to some extent. Females build substantially larger cocoons than males, regardless of host type. Male cocoons on deciduous trees such as the black locust are generally adorned with leaf fragments, while female cocoons are typically covered with stems or petioles (see figure on page 67). Males living on black locust frequently hang their cocoons on petioles, while females seem to prefer substrates of thicker diameter. Males inhabiting red cedar are known to frequent the lower branches of the tree, while females occur higher up. These differences suggest habitat partitioning: males use lower and more peripheral branches, while females use the higher and more central branches. This may reduce competition for food, and has the important consequence of placing females on high, large, stable substrates where they are surrounded by strong reinforcing materials with which they can build sturdy egg-protectors. Male cocoons do not need to last the winter and can be built on thin branches or petioles (which may abscise) close to the ground.

Cases made by male and female evergreen bagworms prior to pupation on black locust trees. Case adornment is not always diagnostic of sex. The pupal exuvium protruding from the bottom of the upper case indicates that it was built by a male. The female's case would contain a caterpillar-shaped pupal cast completely filled with eggs.

METHODS

We will quantify various aspects of cocoon construction and location for bagworms of each sex. The actual measurements taken will depend on the tree species hosting the moths and on the length of the study. The ideal host species is deciduous and has compound leaves (e.g., any locust, ash, or mountain ash). Alternatively, any conifer is suitable. While bagworms can be studied at almost any time of year, most of this analysis can be conveniently done in winter, using overwintering (female) and abandoned (male) cases.

After deciding which variables should be measured, sample approximately 200 cocoons, recording measures for each variable. If the tree is small and the bagworm population is limited, collect the entire population. Quantities that should be measured include:

Height: the height of the cocoon above the ground.

Cocoon width: the greatest diameter of the cocoon.

Cocoon length: the maximum length of the cocoon.

Distance from trunk: the horizontal distance between the cocoon and main trunk (either in a straight line or "as the caterpillar crawls").

Adornment: the amount (%) of the cocoon covered with plant materials. On trees with compound leaves this is usually either leaf fragments or petiole lengths. A convenient method of estimating adornment is to classify cocoons in categories of 25%, e.g., 1-25%, 26-50%, 51-75%, or 76-100%.

Substrate: if the colony is on a deciduous tree, identify substrate types as petioles, this year's growth, or the previous year's growth.

Substrate diameter: the diameter of the twig from which the cocoon was hanging.

Sex: sex of the cocoon builder, as determined after breaking open the cocoon and examining the contents. Female pupae are caterpillar-shaped, have no evidence of wings, and are filled with small, round eggs. Male pupae are much smaller and have clearly visible wings. Larvae cannot be sexed.

Movement: movement patterns may be evaluated in studies of longer duration.

Mark individuals using numbered fiberboard tags (1 × 1 cm) sewed to the cocoon with thread. Mark the original capture site for each individual with a numbered length of flagging. Resample the population at intervals of 2-7 days, and record the distances between successive captures (minimum distance moved). After pupation, sex the individuals that have been followed.

ANALYSES

We are obviously interested in differences between the sexes. For those variables that are continuous (e.g., height, distance) we want to know whether there are differences in location or dispersion. For those variables that are discrete (e.g., substrate types, adornment categories) we want to know whether the categories are independent of sex.

Suggested null hypotheses and statistical tests include:

a. Mean (median) heights were equal for males and females: Student's t test (or Mann-Whitney U test).
b. Mean (median) distances were equal for the sexes: Student's t test (or Mann-Whitney U test).
c. Mean cocoon sizes (areas, volumes) were equal for the sexes: Student's t test (or Mann-Whitney U test). (It may be most convenient to assume that cocoons were cylindrical for this analysis.)
d. Mean (median) substrate diameters were equal for the sexes: Student's t test (or Mann-Whitney U test).
e. Mean (median) distances moved were equivalent for the sexes: Student's t test (or Mann-Whitney U test).
f. Variances for the characters mentioned in a-e above were equivalent for the sexes: F test.
g. Adornment categories were independent of the sex of the builder: Chi-square test, contingency table.
h. Substrate types were independent of the sex of the builder: Chi-square test, contingency table.
i. Substrate diameter was not a function of cocoon size: Linear regression.

Suggested graphical analyses include:

a. Frequency histograms of adornment categories for each sex.
b. Frequency histograms of substrate categories for each sex.

INTERPRETATION

This set of experiments is designed to stimulate your thoughts about sexual differences in behavior. Think about the following questions when interpreting your analyses: How did the sexes differ? What do each of the sexes use their larval cases for, and how might differences in use account for differences in size, strength, adornment, and placement? Why would female bagworms benefit from placing their eggs high in a tree? What fates might befall bagworm eggs that fell to the ground? What choice mechanisms (in the sense explored in Exercise 1 on mechanistic approaches to behavior) can you postulate to account for the distribution of larval cases that you found? Is sexual dimorphism in behavior widespread, or is it limited to specialized lifestyles like the bagworm?

SUGGESTED REFERENCES

Feduccia, A. and B.H. Slaughter. 1974. Sexual dimorphism in skates (Rajidae) and its possible role in differential niche utilization. *Evolution* 28: 164-168.

Gross, S.W. and R.S. Fritz. 1982. Differential stratification, movement and parasitism of sexes of the bagworm, *Thyridopteryx ephemeraeformis*, on red cedar. *Ecol. Entomol.* 7: 149-154.

Horn, D.J. and R.F. Sheppard. 1979. Sex ratio, pupal parasites, and predation in two declining populations of the bagworm *Thyridopteryx ephemeraeformis* Haworth (Lepidoptera: Psychidae). *Ecol. Entomol.* 4: 259-265.

Kaufmann, T. 1968. Observations on the biology and behavior of the evergreen bagworm moth, *Thyridopteryx ephemeraeformis* (Lepidoptera: Psychidae). *Ann. Entomol. Soc. Amer.* 61: 38-44.

Meagher, T.R. 1980. Population biology of *Chamaelirium luteum*, a dioecious lily. I. Spatial distribution of males and females. *Evolution* 35: 1127-1137.

Schoener, T.W. 1968. The *Anolis* lizards of Bimini: Resource partitioning in a complex fauna. *Ecology* 49: 704-726.

Selander, R.K. 1972. Sexual selection and dimorphism in birds. In *Sexual Selection and the Descent of Man*, B. Campbell (ed.). Aldine, Chicago.

Williamson, P. 1971. Feeding ecology of the red-eyed vireo (*Vireo olivaceus*) and associated foliage gleaning birds. *Ecol. Monogr.* 41: 129-152.

18. *The Color, Size, and Spacing of Eggs*

ABSTRACT

The rates of predation on eggs that differ in color and size will be investigated in two different habitats.

INTRODUCTION

A variety of explanations have been given for the variation in egg color among ground nesting birds. An egg of a particular color may reduce mortality because it is visually less conspicuous. Less cryptically colored eggs may be distasteful; hence, palatability may influence egg color. It has also been suggested that the pigments within the shell serve to filter the light transmitted through the shell, and different wavelengths of light have remarkably different effects on the development of chicks. There also remains the possibility that, because the female covers the eggs during incubation, the color of eggs is under no important degree of selection. This seems unlikely. It has been shown that eggs of gulls are cryptic to corvid predators (Tinbergen et al., 1967), and that the female gulls are careful to remove shells from the nest after the young hatch. Since the inside of the shell is white, the shell fragments are easily identified by the ravens and they search the vicinity of the shells for other eggs. Thus, both egg color and distance between nests influence egg survival (at least in certain species).

Regardless of the cause, egg colors vary among birds and across nesting habitats (see table below). Among North American bird species (including burrowing species), ground nesting species tend to produce buff or brown eggs more commonly than do nonground nesters. Buff or brown eggs are extraordinarily common among ground nesters, and this fact lends support to the hypothesis that those colors are harder for visual predators to detect.

Variation in the color of eggs among nonground nesting and ground nesting birds in North America.

EGG COLOR	NONGROUND NESTERS	GROUND NESTERS
White-Gray	245	163
Blue-Green	58	19
Buff-Brown	23	116

(From Reed, 1965.)

In addition, blue or green eggs are rare among ground nesting species. We should be aware that the data only indicate differences in relative abundances of color types, and do not indicate that any particular color is exclusively found among ground nesters.

Not all ground nesting species nest in the same habitats (see table below). For simplicity we have divided ground nesters into those that nest in forested areas and those that nest in more open habitats.

Variation in egg color among ground nesting birds in different habitats.

EGG COLOR	FOREST	OPEN
White-Gray	34	129
Blue-Green	3	16
Buff-Brown	5	111

Again the patterns are nonrandom. Species that nest in the open frequently lay buff-brown eggs, whereas those species that nest in forests generally lay white eggs. It should be noted that the pattern of color variation among forest dwelling ground nesters is identical to that on nonground nesters. It would appear that if egg color is under selection, then the pattern of selection is different in different habitats.

Birds that are ground nesters can be divided into two categories: passerines and nonpasserines (see table below). The former are the songbirds, and the latter are all other birds, including the gallinaceous fowl (chickens, quail, and grouse) and anatids (ducks)—two large groups that are generally ground nesters.

Variation in egg color among passerine and nonpasserine species in different habitats.

	FOREST		OPEN	
EGG COLOR	PASSERINE	NONPASSERINE	PASSERINE	NONPASSERINE
White-Gray	28	6	55	74
Blue-Green	3	0	4	12
Buff-Brown	0	5	0	111

Several patterns emerge in the above table. First, it is uncommon for nonpasserines to nest in forested areas. Second, passerines that nest on the ground do not lay buff or brown eggs. Since the data are separated along taxonomic lines it may be that the observed patterns are a consequence of the phylogeny of each lineage, and the variation in color may or may not have any particular survival value. However, since passerines are generally smaller than nonpasserines, it may well be the case that the variability in egg color is associated with egg size. Egg size is a variable that we can manipulate, whereas it is difficult to juggle phylogenies.

METHODS

We will deal with two colors of eggs, two sizes of eggs, and two habitats. We will dye eggs either brown or white (regardless of their original color). We can obtain small chicken eggs for the large eggs and quail eggs for the small eggs. We will spray our eggs with clear Krylon™ to make their odor more uniform.

The eggs will be placed in grids of 9 eggs each (see figure: You may choose another pattern; our design follows Tinbergen et al., 1967). Ten grids will be established in a woodlot, and in an adjacent field. Eggs will be placed 2 m apart. Allow a space of 10 m so that the eggs within a grid are always closer to grid

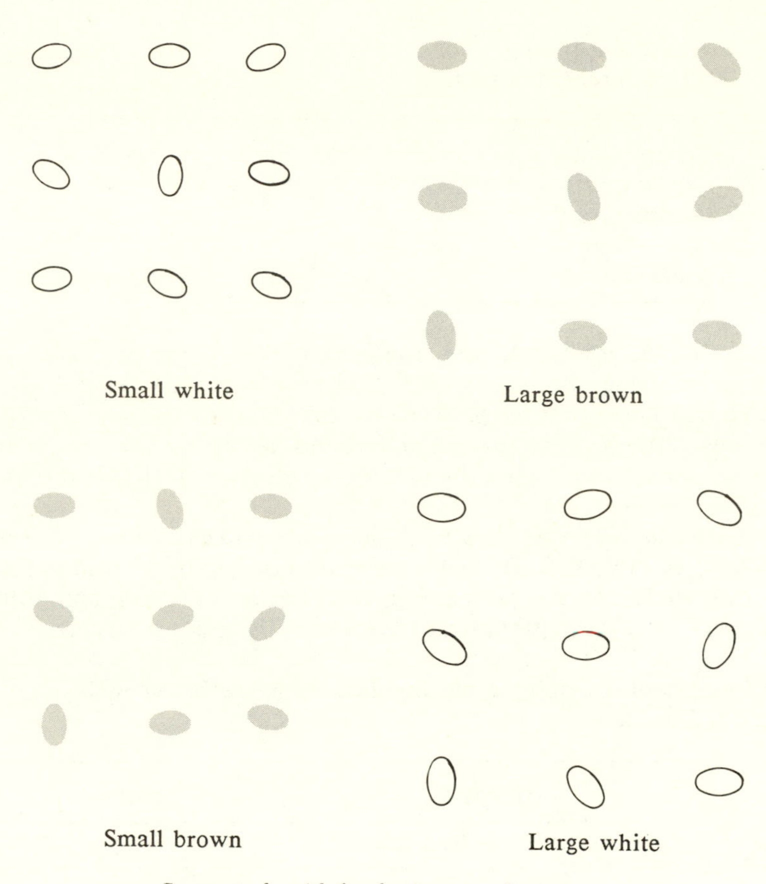

Suggested grid for laying out the eggs.

members than to other grids. Using a regular spacing pattern allows you to relocate eggs more easily. It may be useful to mark each egg with a small stick (cotton swabs work fine) as vegetational growth in the spring may quickly change both types of habitat and hinder relocation of the eggs. The grids should be checked daily to determine the number of eggs lost. Note whether eggs were removed completely or broken open (suggesting terrestrial vs. avian predators).

ANALYSES

Initially we used equal numbers of eggs of each kind. We divided them evenly between forest and field and followed the course of egg loss. When all of the eggs were gone the losses were identical by size and habitat. That is, differences due to size, color or habitat appeared as transient effects in the data. What we conclude from the study may therefore depend on when we sampled.

The data may be summarized in either of two ways. You may examine the number of eggs in each grid when, say 50% of *all* eggs have been taken. Alternatively, if there are great differences between habitats in the overall rate of egg loss, then you may wish to examine each habitat at different times. If you summarize the data in this way, then you have counts of the number of surviving eggs in each treatment. These data can be analyzed with a multiple G test.

Suggested null hypotheses include:

All of the following use the multidimensional contingency table (*G* test).

a. Survivorship is independent of egg color (or size, or habitat). These are separate tests for primary effects.
b. There are no significant interactions between egg color and egg size (egg color and habitat, egg size and habitat). These hypotheses deal with interactions among pairs of variables.
c. There is no interaction among egg color, egg size or habitat. This hypothesis deals with interactions among all three variables.

INTERPRETATION

This set of experiments is designed to stimulate your thoughts about the selective advantages of egg color, egg size, and nest dispersion in birds that nest on the ground. Think about the following questions when you interpret your analyses: What were the primary effects of color, size, and dispersion? (Put another way, if you were a ground nesting bird, what color and size would you want your eggs to be, and how close would you want your neighbors to be?) What do the various possible interactions among primary effects mean in a biological sense? Could you determine differences in the preferences of different types of predator? What selective forces other than predation may determine egg size and shape? Why aren't all eggs the same color and shape?

SUGGESTED REFERENCES

Coleman, M. and G.R. McDaniel. 1975. The effect of light and specific gravity on embryo weight and embryonic mortality. *Poult. Sci.* 54: 1415-1421.

Cott, H.B. 1952. The palatability of the eggs of birds. *Proc. Zool. Soc. Lond.* 122: 1-54.

Lack, D. 1958. The significance of the color of turdine eggs. *Ibis* 100: 145-166.

Nicolaus, L.K. et al. 1983. Taste aversion condition of crows to control predation of eggs. *Science* 220: 212-214.

Pettingill, O.S. 1970. *Ornithology in Laboratory and Field*. Burgess,Minneapolis, MN.

Reed, C.A. 1965. *North American Bird Eggs*. Dover, New York.

Tinbergen, N., M. Impekoven and F. Franck. 1967. An experiment on spacing-out as a defense against predation. *Behaviour* 28: 307-321.

NOTES ON THE STUDY ORGANISMS

This experiment is most successful when conducted in the spring or summer. Predators frequently ignore eggs in autumn, a time when naturally-occurring eggs would be unavailable. It may be possible to deduce something about the species of predator from the patterns of egg destruction: avian predators tend to eat the eggs where they find them; skunks, rats, raccoons, and other mammalian predators may carry the eggs off.

19. The Structure of Human Groups

ABSTRACT

Inter- and intrasexual variation in the composition of groups of animals is examined by observing temporary, informal groups of humans.

INTRODUCTION

The dispersion, or arrangement, of animals in space and time is an obvious feature of populations. In some instances, individuals are solitary and exclude conspecifics or other competitors from their territories. This is the case with many species of songbirds during the breeding season. At other times, or in other species, individuals may congregate in large groups, as is the case when songbirds form flocks over winter, or when fish shoal or school.

The spatial structure of a population is often complex. Sea bird colonies, for example, may form impressive local concentrations of birds. Although clumped or aggregated when viewed on a large scale, individuals within the colonies are usually very evenly spaced, and may, in fact, defend small sites or territories within the larger aggregation. The change is one of the scale on which the dispersion pattern is studied. When the colony is viewed as a whole, the birds are clearly clumped together. Within the colony, however, the birds are uniformly spaced and territorial.

There are some useful generalizations that can be drawn concerning variation in dispersion patterns. Given a suitable scale for examination, social animals usually show some variation of clumped dispersion. Highly mobile animals may vary greatly in the sizes and sexual compositions of their clumps or groups at different times. In some cases, there appears to be clear adaptive significance to this variation. Among African antelopes, for example, small, secretive species usually occur as a monogamous family unit in a well-defended territory. At the other extreme, large, obvious species such as the buffalo are usually found in nomadic, polygamous herds. These patterns are attributable to differences in feeding strategies and predator avoidance, both of which depend on body size. Similar patterns have been identified in carnivores and primates.

The analyses we will perform could be easily adapted to studies of virtually any animal that forms groups, but in this exercise we will consider variation in group composition in the locally abundant, free-ranging primate, *Homo sapiens*. In an elegant monograph, Joel Cohen (1971) explored variation in and among various primate groups. He provided a protocol for data collection and analysis that we will simplify in this study. Cohen cautioned that although patterns may be found in group structure and composition, in themselves such patterns provide no evidence of causal mechanisms. While this is a general caution, it is especially important when we deal with our own species. The motives, emotions, and desires that influence our own actions may or may not explain patterns that we see when we look at others. The patterns themselves do not provide an explanation. They simply describe the ways in which groups are formed; i.e., the social structure of the population. This description is obviously an important part of any complete analysis of animal social behavior.

METHODS

The data needed for this analysis are the sizes and sexual compositions of human groups. These can be recorded by solitary or paired observers watching informal (unorganized) groups that occur under similar circumstances. Examples of such groups might be those entering (or leaving) the library, rathskeller, bank, or church; those sitting at library or study tables; those arriving in cars, etc. Because the circumstances surrounding each of these types of groups are likely to be different, data can be pooled only for those observers watching the same type of group. If each observer watches a different type of group, each data set must be analyzed separately.

Data can be rapidly and unobtrusively recorded in tabular form (see figure). Record only those groups that enter a sample area that is defined before observation begins. Suitable sample areas include such things as an entranceway, a cross-section of sidewalk or road, library tables of equal size, etc. Precise definition of the sample area is necessary to avoid biasing the sample by the investigator. Most humans habituate rapidly to an observer's presence, so a single watcher sitting on a bench or the grass quietly making notes and looking thoughtful will probably not disturb natural group formation (under most circumstances). As the analyses we will perform require fairly large sample sizes, the

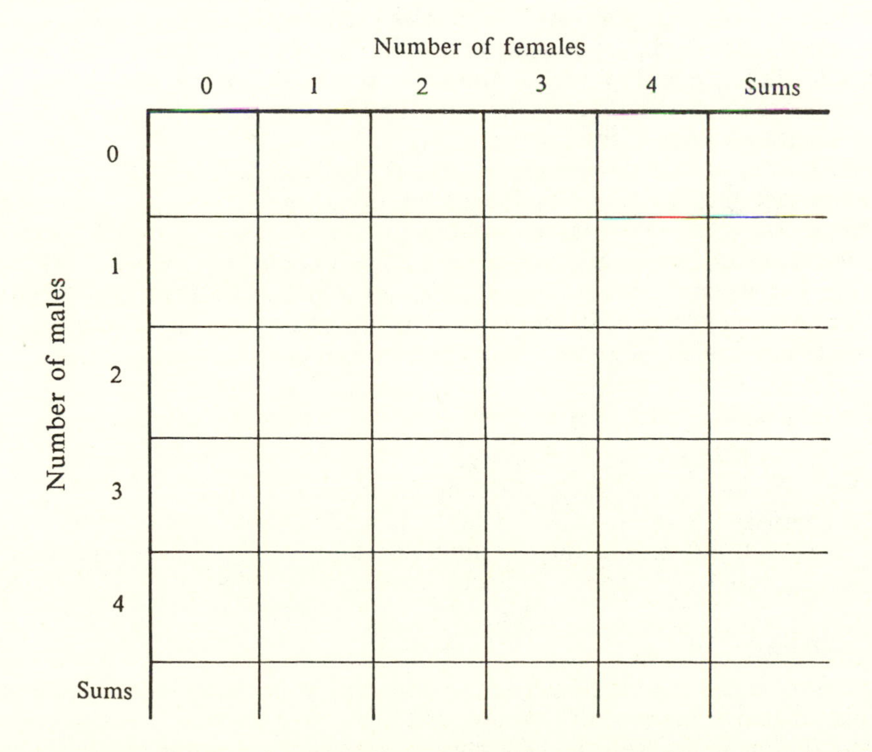

The matrix used to collect and organize data. Groups of as many as 8 persons can be included here, but the table can be enlarged as appropriate. Groups of only males or only females are tabulated in the first column or row respectively. Groups of mixed sexual composition are noted in the body of the table. Row totals will indicate the distribution of males without regard to females. Column totals will indicate the distribution of females without regard to males.

more groups that are observed the better. We suggest that all investigators observe the same type of group so that several hundred groups can be recorded quickly.

ANALYSES

The data allow us to describe the social structure for our population in several ways, including average group size, variability in group size, range in sizes, etc. We are primarily interested, however, in inferences about the interactions between and within the sexes. Posed as questions, these inferences ask: Did males and females assort independently of one another? Were individuals of the same sex randomly distributed or did they avoid one another or clump together? These questions can be answered through comparisons to a binomial and Poisson distribution, respectively.

Suggested null hypotheses and statistical tests include:

a. Distribution of the sexes within groups of equal size was binomial: Chi-square test, analysis of binomial distributions. This is equivalent to asking the question of independent assortment of the sexes. Note (as described in the section on binomial analyses) that groups of equal size must be analyzed separately, thus total sample sizes will have to be large.
b. Distribution of individuals of the same sex was random: Chi-square test, comparison to a Poisson distribution. Use the row totals (i.e., the distribution of males independent of females) and column totals (i.e., the distribution of females independent of males) from the matrix shown on page 75 for your observed distributions. Since there will not be any zero class in most cases, the truncated Poisson must be used.
c. Distribution of males was independent of that of females in the population: Chi-square test, contingency table analysis. This analysis can only be performed for data sets that include the number of observations having zero individuals. It is thus very limited in application, but will be possible in some cases. For example, some library tables, or tables at McDonald's, or church pews, etc., may have been unoccupied, yielding the zero class. Note that each sample unit (table, pew, etc.) must be of equal size for this analysis.

Suggested graphical analyses include:

a. Frequency histograms of group sizes.
b. Frequency histograms of the distribution of sexes within groups of the same size compared to those expected for random (binomial) distribution.
c. Frequency histograms of the distributions observed for each sex compared to those expected for random (Poisson) distributions.

INTERPRETATION

This set of experiments is designed to stimulate your thoughts about the ways animals structure their social groups and the ways in which the sexes may interact. Think about the following questions when you interpret your analyses: Were males and females behaving independently of each other? Did group size affect the relationship among the sexes? Was there any evidence that members of the same sex grouped together or, alternatively, avoided one another? What do these findings tell you about the ways that individuals relate to members of the same and the opposite sex? If you compared groups entering and leaving a building, did you find any differences between them? Can you propose explanations for these differences? Are these explanations supported by the data? What further experiments would be necessary to provide further explanations?

SUGGESTED REFERENCES

Cohen, J. 1971. *Casual Groups of Monkeys and Men.* Harvard Univ. Press, Cambridge, MA.

Crook, J.H. 1970. The socio-ecology of primates. In *Social Behaviour in Birds and Mammals*, J.H. Crook (ed.). Academic Press, London.

Crook, J.H. 1970. Social behavior in ethology. In *Social Behaviour in Birds and Mammals*, J.H. Crook (ed.). Academic Press, London.

Crook, J.H. and R.P. Michael. 1971. Primates and human ethology. In *Comparative Ethology and Behavior of Primates.* Academic Press, London.

Davies, N.B. and J.R. Krebs. 1978. Introduction: Ecology, natural selection and social behaviour. In *Behavioural Ecology*, 1st ed., J.R. Krebs and N.B. Davies (eds.). Sinauer, Sunderland, MA.

Downhower, J.F. and L. Brown. 1979. Seasonal changes in the social structure of a mottled sculpin (*Cottus bairdi*) population. *Anim. Behav.* 27: 451-458.

Jarman, P.J. 1974. The social organisation of antelope in relation to their ecology. *Behaviour* 48: 215-267.

Kleiman, D.G. and J.F. Eisenberg. 1973. Comparisons of canid and felid social systems from an evolutionary perspective. *Anim. Behav.* 21: 637-659.

Lack, D. 1968. *Ecological Adaptations for Breeding in Birds.* Methuen, London.

Pulliam, H.R. and T. Caraco. 1984. Living in groups: Is there an optimal group size? In *Behavioural Ecology*, 2nd ed., J.R. Krebs and N.B. Davies (eds.). Sinauer, Sunderland, MA.

20. Oviposition by Bean Beetles

ABSTRACT

Oviposition site preference of female bean beetles are examined by allowing females to choose among sites with different qualities. The consequences of oviposition site choice are evaluated by determining eclosion success for each site type.

INTRODUCTION

There are many occasions in an animal's life that require some type of choice: the animal must select one option from some set of possibilities and reject the other options. Some choices, such as accepting or rejecting a particular food item, may not have immediate consequences for an animal's fitness. Other choices, such as accepting or rejecting a mate, may have extremely important consequences. We might expect that the more important a particular choice is to an individual's fitness, the more intensely selection has refined the choice process.

Oviposition site choice can have important consequences for individual fitness in many species. Such choices affect exposure to predators, weather, and critical resources such as water or food. The larvae of many insects, for example, cannot disperse away from their oviposition site. Larvae are constrained to the site chosen by their mother; hence oviposition site choice determines larval survival to a great degree. The bean beetle *Callosobruchus maculatus* is such an insect. The larvae of this bruchid beetle develop at their oviposition site, and sites of differing qualities affect larval survival, growth rates, and subsequent adult fecundity and longevity. In this exercise we will examine oviposition site preferences of females and evaluate the consequences of these preferences to larvae and, subsequently, to female fitness.

THE STUDY ORGANISMS

Females of the spotted bean beetle *Callosobruchus maculatus* lay their eggs singly on the surface of legume seeds. After 8-10 days, the larva hatches and chews into the bean. It completes its development in the bean and gnaws a chamber in the cotyledons, where pupation and eclosion occur. The adult emerges through an exit hole in the seed coat approximately 34 days after oviposition. Adults do not need to eat, although longevity and fecundity improve somewhat if water or sugar solution are available.

Several studies, most notably those of Mitchell (1975, 1983) have demonstrated that larval survival and adult life history parameters are determined by the quality of the bean in which the larva grows. Since larvae cannot move among beans, they must use those resources chosen by their mothers. Since adults do not normally feed, female choice of a particular oviposition site determines both larval survival and the fecundity of daughters. This is a most important choice. Many types of legumes are not suitable for larval growth, either because of chemical protection or thick seed coats, and even those beans suitable for larval growth are highly variable in size. For example, one common food of this beetle is the mung bean *Phaseolus aureus*. Many populations of mung beans have a high frequency of beans too small to support even a single larva. Larval survival and adult size are directly and linearly proportional to bean size when only one larva grows in a bean. When more than one larva is present, interactions between them greatly reduce survival, and eliminate the dependency on bean size.

METHODS

Analysis of oviposition site choice and its consequences will involve several independent experiments. Each of these will be started during one lab period, and data on survival (if desired) will be collected approximately one month later. Fecundity data must be collected after adults emerge.

Oviposition site preferences among mung beans

Fill a total of 25 petri dishes each with exactly 10, 20, 30, 40, and 50 mung beans. Use an aspirator to remove 1 newly-eclosed female bean beetle (see figure) from the stock culture, and place her with two males in a dish. Seal the dish with tape, and label with the experiment name, your name, and the date. Dishes will be cultured at approximately 20° C for one week to allow the females to oviposit. At that time, the beans in each dish should be removed and beans sorted into groups having 0, 1, 2, 3, etc. eggs per bean. The frequency of each class of beans and the total number of eggs laid by each female in each dish should be recorded. Place the pooled beans in suitable plastic containers and resume culturing until adults emerge. Determine survival to adulthood by counting the number of adults produced for each class of beans.

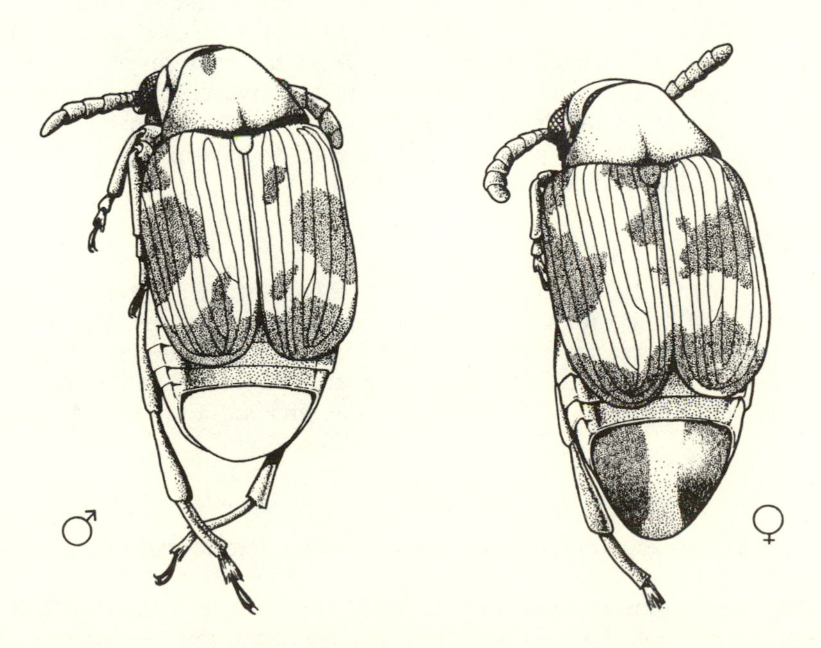

Dorsal views of anaesthetized male and female bean beetles, Callosobruchus maculatus. *Note the plate covering the posterior abdomen (pygidium). The female's pygidium is enlarged and darkly pigmented on the sides.*

Oviposition site preference among beans of different species

Fill at least 25 petri dishes with a mixture of 10 beans each of the following types: garbanzo (*Cicer arietinum*); mung (*Phaseolus aureus*); kidney, pinto, navy, or black (*P. vulgaris*); lima (*P. limensis*); soy (*Glycine max*); lentils (*Lens culinaris*); and black-eyed peas (*Vigna unguiculata*). Place a single newly-eclosed female and at least two males in each dish. Seal the dishes and label appropriately. Culture the dishes for approximately a week to allow oviposition, and remove all beans, counting the numbers of eggs on each type. Choose at least 10 beans of each type from the stock containers and measure their lengths and widths using calipers. This will allow calculation of the surface areas (L × W) of each type.

Consequences of preferences for particular bean types

Some of the bean types used in the analyses of oviposition preferences will not have been chosen at all. Analyses of the consequences of bean choice require that females be forced to use all types. Consequently, fill at least 10 petri dishes with 100 beans of each of the beans used previously (one type of bean per dish). Note that some beans will be so large that plastic containers other than petri dishes may be necessary. Place a single newly-eclosed female and 2-3 males in each dish, seal, and label. Sort through the beans regularly during the next week, and remove all beans with single eggs. Pool these according to type so that large numbers of each type having solitary eggs are made. Culture the pooled collections until adults emerge and determine survival to adulthood by counting adults produced from each bean type. Notice that this procedure eliminates larval interactions and ensures that results reflect bean quality, not larval competition.

Consequences of bean quality to adult fecundity

For each bean type that produced adult females in the last experiment, fill 10 petri dishes with 200 mung beans. Place a single newly-eclosed female and 2-3 males in each dish, using individuals derived from the analyses of consequences given above. Culture these dishes until the females have died (not more than 30 days). Remove all beans and determine the fecundity of each female by counting the total number of eggs she produced. Place each female in a small vial, label, and dry at 50° C for 24 hours before weighing.

ANALYSES

We are interested in determining the relationships between female fecundity and the availability of oviposition sites. In addition, we are interested in evaluating female preferences when given choices among beans of a single species, and when confronted with beans of different species and strains. Finally, we want to evaluate the consequences of this choice to the survival and fecundities of the offspring produced.

Suggested null hypotheses and statistical analyses include:

a. Average female fecundity on mung means was not determined by bean number: ANOVA.
b. Oviposition on mung beans was random: Analyses of dispersion using the Poisson distribution. (Do not pool data. Analyze each female's pattern separately. This is necessary since Poisson frequencies are dependent on the distributions mean, and each female will have a different fecundity yielding a different mean.)

c. Oviposition on beans of different types was random: Chi-square test, a priori hypotheses that the number of eggs per bean type is directly proportional to the surface area available for that type.
d. Survival to eclosion was equivalent for mung beans having different numbers of eggs per bean: Chi-square test, equal proportions of adults for each treatment.
e. Survival to eclosion was equal for all bean types: Chi-square test, a priori hypothesis that survival was equivalent for all bean types. Notice that a Student's *t* test might be appropriate if survivors only emerged from two types, which is a possibility.
f. Female fecundity was equal for females raised on each bean type: Analysis of variance (or Student's *t* test) comparing mean fecundities for each type.

Suggested graphical analyses include:

a. Frequency histograms comparing observed and expected frequencies for beans having different number of eggs.
b. Scatter plot for bean weight vs. number of survivors for mung beans. If enough classes are available, correlation coefficients might be included.

INTERPRETATION

This set of experiments is designed to stimulate your thoughts about the ways that animals make choices and the consequences of those choices to individual fitness. Think about the following questions when interpreting your analyses: Did females adjust their egg production in response to changes in the availability of oviposition sites? Did females count the number of oviposition sites available to them? Did females count the number of eggs already present at each oviposition site? Did females evaluate the type of bean before ovipositing on it? What were the consequences of laying different numbers of eggs on beans of the same type? What were the consequences of laying eggs on different types of beans? Could a female determine the number of daughters she left by being choosy or would random oviposition leave the same number on average? Could a female determine the number of granddaughters she left by being choosy? What are the selective forces that favor the evolution of this type of choice (i.e., would you expect choosiness of this type to be found in all populations)?

SUGGESTED REFERENCES

Avidov, Z., S.W. Applebaum and M.J. Berlinger. 1965. Physiological aspects of host specificity in the Bruchidae: II. Ovipositional preference and behavior of *Callosobruchus chinensis* L. *Entomol. Exp. Appl.* 8: 96-106.

Avidov, Z., S.W. Applebaum and M.J. Berlinger. 1975. Physiological aspects of host specificity in the Bruchidae: III. Effect of curvature of surface area on oviposition of *Callosobruchus chinensis* L. *Anim. Behav.* 13: 178-180.

Janardan, K.G., H.W. Kerster and D.J. Schaeffer. 1979. Biological applications of the Lagrangian Poisson distribution. *BioScience* 29: 599-602.

Mitchell, R. 1975. The evolution of oviposition tactics in the bean weevil, *Callosobruchus maculatus* F. *Ecology* 56: 696-702.

Mitchell, R. 1980. Lagrangian Poisson model. *BioScience* 30: 288.

Mitchell, R. 1983. Effects of host plant variability on the fitness of sedentary herbivorous insects. In *Variable Plants and Herbivores in Natural and Managed*

Systems, R.S. Denno and M.S. McClure (eds.). Academic Press, New York.
Wasserman, S.S. and D.J. Futuyma. 1981. Evolution of host plant utilization in laboratory populations of the southern cowpea weevil, *Callosobruchus maculatus* Fabrivius (Coleoptera: Bruchidae). *Evolution* 35: 605-617.
Wright, J.S. 1983. The dispersion of eggs by a Bruchid beetle among Scheela palm seeds and the effect of distance to the parent palm. *Ecology* 64: 1016-1021.

NOTES ON THE STUDY ORGANISMS

Callosobruchus maculatus can be easily cultured in a screen-covered jar or dish filled with mung beans or black-eyed peas. The insect is a common pest of stored beans and is frequently found in beans of the cowpea group wherever beans are stored. Health food stores have provided the authors and others of our acquaintance with many strong cultures. A culture of the beetles can also be obtained by writing to Luther Brown, Department of Biology, George Mason University, Fairfax, VA 22030.

21. Assortative Mating in Soldier Beetles

ABSTRACT

Size-dependent mating patterns of soldier beetles are examined by measuring copulating and single individuals. Behavioral interactions between and within the sexes are used to evaluate the importance of size to social behavior and mate choice.

INTRODUCTION

Sexual selection occurs when some individuals obtain more mates or produce more offspring than other members of the same sex. This process is thus distinct from natural selection, which occurs when individuals differ in their survival ability or their physiological capacity to produce offspring. Like natural selection, sexual selection can occur in several different patterns. For example, if the more extreme individuals (e.g., the largest or the smallest) in a population are more successful in obtaining mates, then directional selection towards that extreme will occur. If the average individuals are more successful than the extremes, then stabilizing selection will occur. If both extreme forms are more successful than the average, then disruptive selection will occur. Finally, the success of any single type of individual may depend on the types of competitors it experiences, thus frequency-dependent selection may occur.

Any of these selection processes is possible whenever mating within a population is nonrandom. Such assortative or nonrandom mating has been demonstrated in several vertebrates and invertebrates, and is probably the rule rather than the exception. In some cases, it may be attributable to simple physical constraints; for example, a small female beetle may not be able to mate with any male much larger than herself. In other cases, assortative mating may be due to either female choice of males possessing particular characteristics, or to interactions among males (or females) that restrict access to females. In this exercise we will examine mating patterns in the cantherid soldier beetle *Chauliognathus pennsylvanicus*. While several other species might be used (e.g., other soldier beetles, Japanese beetles, blister beetles, or lady bird beetles), this species is abundant throughout the eastern United States, and has recently been the subject of several analyses.

THE STUDY ORGANISMS

Chauliognathus pennsylvanicus, the soldier beetle, has recently been the subject of several analyses of assortative mating. In summary, some of these studies have demonstrated directional selection for increased body size by showing that mated individuals are larger than nonmated ones. Other analyses have shown that selection occurs only on some characteristics, including the length of the basal antennal segment in males (a secondary sexual character). Still other analyses have suggested that prolonged copulation may allow individuals to out-compete wasps, which use the same types of pollen used by beetles as a food source. Most studies have shown a positive correlation between the sizes of male and females captured in copula.

Only one study (McCauley and Wade, 1978) has examined behavioral interactions quantitatively. This analysis demonstrated that all encounters between beetles could be characterized as being aggressive. These authors concluded that a type of de facto female choice for large males was in effect: females reject all males, and thus mate only with those males large enough to overcome and subdue them. This choice model (the Kence-Bryant model) results in females mating with males who are larger than themselves.

Despite this attention to the breeding biology of this insect, little is known about its general life history. Some authors maintain that the larvae are predaceous and live in forest leaf litter. Others maintain that the larvae feed on pollen. No one knows where the eggs are laid, and dispersion patterns and survivorship are poorly known at best. All authorities agree that the adults feed on pollen from goldenrod (*Solidago*), *Bidens* spp., and other late-summer flowering plants. The beetles are frequently very abundant, and may reach densities of hundreds per square meter.

METHODS

We will use an early successional field with a high-density soldier beetle population for our study site. Move through the field, collecting solitary and mating individuals as you go. As you capture the beetles, measure the length of the elytra and the width of the thorax as indicated in the figure below.

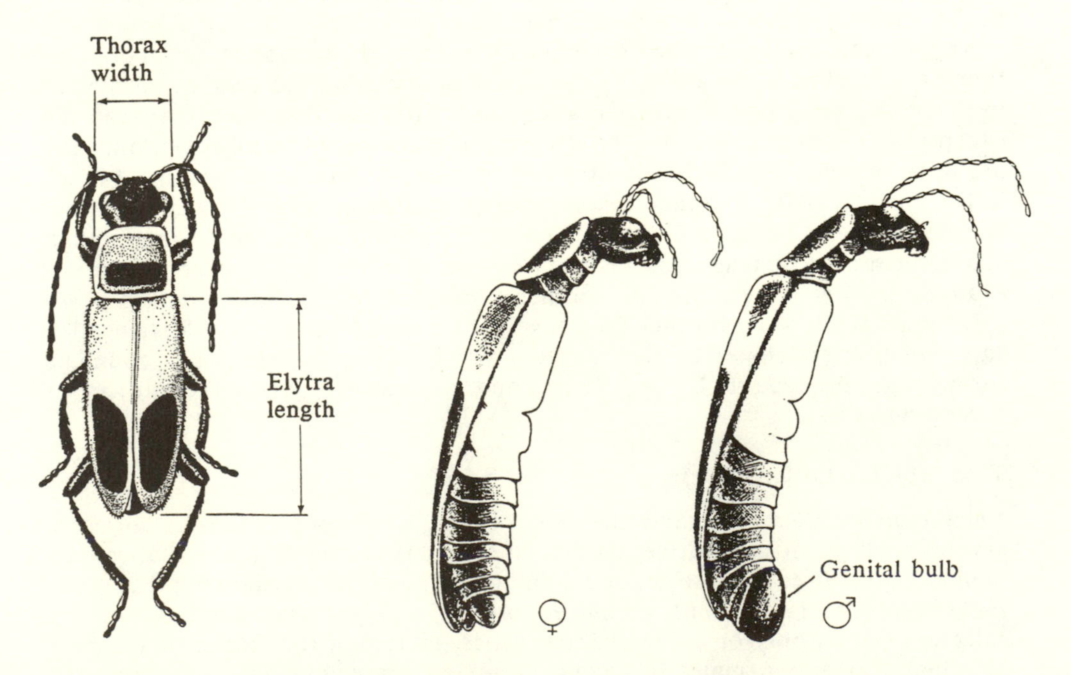

The soldier beetle Chauliognathus pennsylvanicus. *The left-hand illustration shows the beetle's dorsal surface, while the other figures illustrate the external difference between male and female abdomens. Note that legs are not drawn for the sake of clarity.*

Place each individual or pair in a small vial. This will ensure that no beetles are measured more than once and will allow weights to be taken should you so desire. Collect as many beetles as possible. Be sure to include at least 50 copulating pairs and 50 single beetles of each sex. If weights are to be used, return the beetles to the lab, freeze them to kill, and dry them at about 80° C for 24 hours before weighing.

Before leaving the study site, observe the behavioral interactions between beetle dyads. Define the initiator as the individual that approaches another to within 1 cm. Define the recipient as the approached beetle. Record what happened during each encounter (e.g., copulated, fought, etc.), the sex of each individual, and which individual (initiator or recipient) ended the encounter (ran away). Capture the interactors and measure them. Observe as many male-female, male-male, and female-female encounters as possible.

ANALYSES

We are primarily interested in the following questions: Is there any significant correlation between male and female sizes for beetles captured in copula? Were the copulating beetles different from noncopulating ones. What were the behavioral interactions between and within the sexes, and what do they tell us about female choice, male-male competition, etc.?

Suggested null hypotheses and statistical analyses include:

a. There was no significant correlation between male and female sizes for copulating pairs: Pearson or Spearman correlation coefficient. Note that the measures of size will depend on the data collected and might include elytra length, thorax width, body surface area (elytra × thorax), weight, etc.
b. The slope of the line relating the sizes of copulation males and females is not different from zero: Linear regression analysis.
c. The average sizes of copulating males (females) were not different from those for single males (females): Mann-Whitney U test for comparing medians or the Student's t test for comparing means.
d. Mating individuals were not more or less variable than nonmating ones: F test, comparison of ranges.
e. Males in copulating pairs were not larger than their mates: Mann-Whitney U test for comparing medians; Student's t test for means.
f. Winners of behavioral interactions were not larger than losers: Mann-Whitney U test for medians; Student's t test for means.
g. Success in winning an encounter is independent of sex of the recipient: Chi-square test, contingency table.

Suggested graphical analyses include:

a. A scatter plot, with regression line, of the sizes of male and females captured in copula.
b. Frequency histograms of the sizes of mating males (females) and of nonmating males (females). Note that these could be plotted on the same set of axes, keeping the two sexes on separate figures.

INTERPRETATION

This set of experiments is designed to stimulate your thoughts about sexual selection and nonrandom mating. Think about the following questions when interpreting your analyses: What does a correlation between the sizes of copulating males and females mean? How large was the correlation, and how does it compare with those reported in other studies? What would it mean if you found that body areas were correlated but body lengths were not? Is a significant correlation of sizes sufficient evidence of sexual selection? What would it mean if mating individuals were larger than nonmating beetles? What would it mean if mating and nonmating individuals had the same variability (or different variabilities)? What does it mean if females are larger on average than males? Did you find any evidence indicating beetles that won an aggressive encounter were different from those that lost? Is assortative mating a necessary consequence of sexual selection? Soldier beetles are ectotherms and are thus more active late in the day and on sunny days. How do you think that temperature and time of day might affect the patterns you found? If you did find evidence of directional selection, how do you explain the presence of "unsuccessful" or nonmating individuals? If you found evidence of stabilizing selection, how do you explain the observed variability in the population? If you found no evidence of selection, how do you explain your results in light of the other studies that have reported selection?

SUGGESTED REFERENCES

Bateson, P. 1983. *Mate Choice*. Cambridge Univ. Press, Cambridge.

Mason, L.G. 1964. Stabilizing selection for mating fitness in natural populations of *Tetraopes*. *Evolution* 18: 492-497.

Mason, L.G. 1980. Sexual selection and the evolution of pair bonding in soldier beetles. *Evolution* 34: 174-180.

McCauley, D.E. 1979. Geographic variation in body size and its relation to the mating structure of *Tetraopes* populations. *Heredity* 42: 143-148.

McCauley, D.E. 1980. Application of the Kence-Bryant model of mating behavior to a natural population of soldier beetles. *Amer. Natur.* 117: 400-402.

McCauley, D.E. and M.J. Wade. 1978. Female choice and the mating structure of a natural population of the soldier beetle, *Chauliognathus pennsylvanicus*. *Evolution* 32: 771-775.

Woodhead, A.P. 1981. Female dry weight and female choice in *Chauliognathus pennsylvanicus*. *Evolution* 35: 192-193.

NOTES ON THE STUDY ORGANISMS

Although this analysis emphasizes soldier beetles, many other animals can answer the same questions. Any species that is abundant and mating, whether it is a beetle, a frog, or a bug, can be examined using the same basic techniques. One particularly nice thing about soldier beetles is the color polymorphism found in many populations. This allows expansion of the analyses proposed here to deal with assortative mating within or between color morphs. John Kochmer, of Yale University, has proposed the following morph types for his own work on gene flow in soldier beetles (see figure on page 87).

1. "Spot" has an elliptical black spot on each elytra. Both upper and lower margins of this spot are smooth.
2. "Spot hybrid" has a rough ellipse with a jagged anterior (humeral) edge. The spot itself may be large or small. The critical factor is the rough anterior edge.

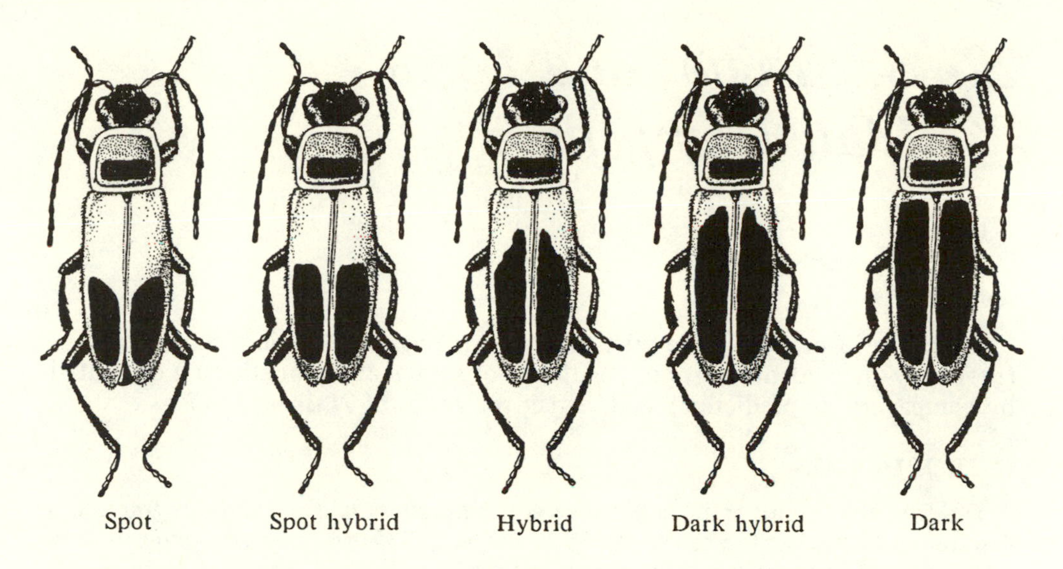

Color morphs of the soldier beetle Chauliognathus pennsylvanicus *as proposed by J. Kochmer.*

3. "Hybrid" has an elliptical spot with an elongated neck running along the inside edge (sutural edge) of each elytra.
4. "Dark hybrid" has almost the entire elytra covered with black, but has pronounced shoulders at the anterior of the dot.
5. "Dark" has completely dark elytra.

Assortative mating by morphs can be analyzed through a contingency table (Chi-square) testing the null hypothesis that the morph of the male was independent of the morph of the female.

If you choose to work on a species other than the soldier beetle described here, keep your eyes open for potential color or pattern polymorphisms. Many animals differ in their spot patterns, melanism, number of stripes, etc.

22. Sex Ratios and Local Mate Competition

ABSTRACT

Sex ratios in broods of the wasp *Nasonia vitripennis* provide a special case of frequency-dependent selection. The theory of local mate competition is evaluated by comparing its predictions with actual sex ratios of *Nasonia* broods.

INTRODUCTION

The fact that many animals produce equal numbers of sons and daughters has fascinated biologists for centuries. R.A. Fisher (1930) provided an evolutionary explanation for equal sex ratios that was based on the fact that all diploid sexual organisms have one mother and one father—i.e., the average mating success of males and females must be equal. This means that parents will realize the greatest fitness if they invest equally in sons and daughters. More specifically, fitness is maximized if the cumulative effort at the end of the period of parental care is equal for both sexes. When male and female offspring cost the same to produce, sex ratios will approximate unity. When one sex is more expensive to produce, equal investment will result in a biased sex ratio favoring the cheaper sex.

Fisher's arguments apply only to panmictic populations. When breeding is structured, either through differential mating successes or spatial organization, equal ratios may not lead to maximal fitness. Several authors have demonstrated that optimal sex ratios may depend on factors as diverse as the physiological status of a female, the social position of a female or her mates, and the tertiary, or population, sex ratio that offspring will experience.

Populations that are spatially structured provide an interesting case in which optimal sex ratios may not be 1:1. Consider, for example, a large population that is subdivided into many mating groups composed of the offspring of one or more females. If sons typically mate with their sisters, who then disperse to found new groups, selection will favor parents who produce an excess of daughters. The actual ratio of daughters to sons will depend on the number of females who found each population. When single females always start populations, females should produce only enough sons to inseminate their daughters. As the number of additional foundresses increases, the optimal sex ratio approaches unity. The timing of reproduction also affects optimal ratios. If females contribute their broods to the population sequentially, and do not know how many additional females will follow them in the sequences, then the optimal sex ratio for the first female will be strongly female biased, that for the next female will be less biased, that for the third even less. Finally, the relative contribution of each female to the total population size will be a determinant of optimal sex ratios. If a first foundress leaves a large brood with many daughters and few sons, and the second foundress leaves a very small brood, her optimal ratio will be 100% sons.

These relationships make up the theory of local mate competition, first proposed by Hamilton (1970) and subsequently expanded and tested by Werren (1980, 1983). Various parasitic wasps make especially good example cases, since one or more females oviposit on a host, females are inseminated by their brothers, and fecund females disperse to start new populations. Under these conditions, the optimal sex ratio for an individual female depends on the size of her brood, and the probability that her sons will compete with the sons of other females. Specifically, Werren (1980) has argued that the fitness of a second female ovipositing on a previously used host will equal

$$\frac{TX_2}{X_1 + TX_2} \; [1 - X_1 + T(1 - X_2)] + T(1 - X_2)$$

where $T = \dfrac{\text{offspring number of the second female}}{\text{offspring number of the first female}}$

X_1 = proportion of sons in the first female's brood

X_2 = proportion of sons in the second female's brood

This function is maximized when

$$X_2 = \frac{(\sqrt{2X_1 (T + 1)} - 2X_1)}{2T}$$

Furthermore, local mate competition theory predicts that females should be sensitive to the density of competing foundresses within a habitat. If wasp densities are very low, then the probability of joint founding is small, and initial foundresses may produce very few (1 or 2) sons per brood. As wasp densities increase, brood sex ratios should become more equitable.

The jewel wasp *Nasonia* (= *Mormionella*) *vitripennis* is an organism with which predictions of local mate competition theory can be tested. In this exercise we will examine sex ratios in the broods of this wasp and evaluate the ability of individual females to optimize their brood sex ratios.

THE STUDY ORGANISMS

Female jewel wasps (*Nasonia vitripennis*) (see figure) oviposit in the pupae of various cycloraphous flies. Wasp development proceeds inside the fly puparium, eventually resulting in the death of the fly. At 29° C, average development time is about 14 days: 2 days as an egg, 6 days as a larva, and 6 days as a pupa. Adults emerge from the fly puparium through an exit hole. Males eclose first and compete for positions near the exit hole. Females mate as they emerge from the puparium. Males mate many times and, since they have vestigial wings, do not disperse. Females mate once and disperse to find new fly pupae. As in most hymenopterans, males are haploid (resulting from unfertilized eggs) and females are diploid. Internal structures appear to allow females to oviposit fertilized or unfertilized eggs, giving females control over the sex ratio of their broods.

Superparasitism of fly pupae does occur in wild populations. The second female attacking a fly pupa can apparently detect the fact that it has already been parasitized and can vary her brood sex ratio in response. When only one female attacks a fly pupae she produces 5-25% sons. Second foundresses may produce up to 100% sons, depending on their brood size. When many foundresses attack single flies, total population sex ratios approach 50%.

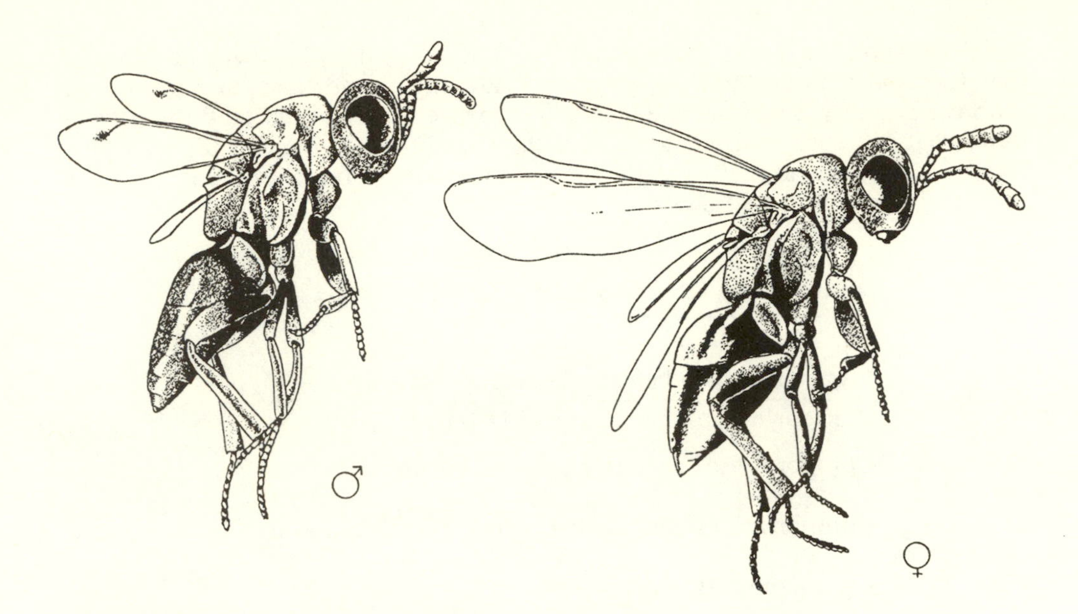

Anaesthetized jewel wasps, Nasonia vitripennis. *The female has a larger body, wider abdomen, and wings that are about as long as her total body. The male is much smaller, has a narrower abdomen, and has wings that are greatly reduced.*

METHODS

Responses of second foundresses

For this analysis we will parallel the methods of Werren (1980), and take advantage of the fact that two genetically homogeneous strains of jewel wasps are commercially available: wild-type (++) and *scarlet eye (ScDr)* mutants. It is safe to assume that all adult females in a culture are mated. Females should be isolated from fly pupae and fed honey for 24 hours prior to use.

Place a single wild type female in a petri dish with four large fly pupae (e.g., *Sarcophaga*). After 24 hours, remove the female and introduce a female *scarlet eye*. Allow her to superparasitize the pupae for 24 hours and remove her. Place the pupae in individual vials, and incubate at room temperature. Count the numbers and phenotypes of males and females emerging at eclosion. Repeat this procedure so that as least 50 pupae are parasitized. For completeness, replication should involve an equal number of trials in which wild-type females follow *scarlet eyes*.

Consequences of population density

For this analysis we will parallel the methods of Werren (1983). Place 4 fly pupae in a petri dish and add 1, 2, 3, 4, 6, 8, or 12 female wasps to the dish. Genotype is unimportant here. Replicate each wasp density at least 5 times. Incubate at room temperature and count the numbers of males and females that emerge at eclosion.

ANALYSES

In the first part of this exercise we are interested in comparing the responses of wasps with the predictions made by local mate competition theory. This is most easily done by calculating the expected sex ratios for second foundresses using the equations given above, and comparing the expectations with the ratios actually observed. Notice that the actual sex ratio for first foundresses must be used to obtain the expected values for the second foundresses. Comparison between observed and expected values should be graphic: plot X_2 as a function of T.

In the second part of the experiment we are interested in the covariance between wasp density and sex ratio. Again, this can be analyzed graphically by plotting sex ratio as function of foundress density.

Since neither relationship expected in these two sets of analyses is linear, linear regression will not be appropriate. Spearman correlation coefficients may be applicable however.

INTERPRETATION

This set of experiments is designed to stimulate your thoughts about the factors that affect sex ratios. Think about the following questions when interpreting your analyses: Why don't jewel wasps produce equal numbers of sons and daughters? What makes jewels wasps different from those organisms that do produce equal numbers of each sex? What other organisms might be expected to show the same kinds of responses found in wasps (think about plants)? How would you expect a wasp to respond to an array of oviposition sites ranging from small (like a housefly) to large (like a fleshfly)? How might a female decide whether or not a fly pupa has already been used by another female? Would the first female to use a pupa want to hide the fact from other females?

SUGGESTED REFERENCES

Fisher, R.A. 1930. *The Genetical Theory of Natural Selection.* Oxford Univ. Press, Oxford.

Hamilton, W.D. 1967. Extraordinary sex ratio. *Science* 156: 477-488.

Werren, J.H. 1980. Sex ratio adaptations to local mate competition in a parasitic wasp. *Science* 208: 1157-1159.

Werren, J.H. 1983. Sex ratio evolution under local mate competition in a parasitic wasp. *Evolution* 37: 116-124.

Whiting, A.R. 1967. The biology of the parasitic wasp *Mormionella vitripennis* (= *Nasonia brevicornis* Walker). *Quart. Rev. Biol.* 42: 333-406.

NOTES ON THE STUDY ORGANISMS

Both wasps and their hosts are available from biological supply houses. They are usually shipped with complete culturing directions, and are very easy to maintain in the lab.

23. *Oviposition Site Choice in Cabbage Butterflies*

ABSTRACT

Oviposition site preferences of cabbage butterflies (*Pieris rapae*) are evaluated by allowing females to lay eggs on several plant species, and on cabbage plants that have been manipulated to simulate leaf damage and prior oviposition. Use of chemical cues during site choice is evaluated by allowing females to lay eggs on paper discs impregnated with plant extracts.

INTRODUCTION

In many species choice of an oviposition site greatly influences reproductive success. This is especially true when the oviposition site is also the site of juvenile growth, as it is for many insects that lay their eggs on specific host plants. Females of such species who lay their eggs on appropriate hosts may ensure their offsprings' survival, while those that oviposit on unsuitable hosts may cause their eggs or larvae to die. Factors determining host plant suitability may be diverse and include such things as species or race, physiological condition, and whether or not it has already been used as an oviposition site by other females. Plants of an inappropriate species or race may be chemically defended against larvae or provide poor nutrition. Plants in poor physiological condition, or those with extensive leaf damage, may provide poor or inadequate diets, and those already supporting the offspring of other females may be overcrowded with competitors or potentially cannibalistic predators.

THE STUDY ORGANISMS

The cabbage butterfly *Pieris rapae* is an animal that demonstrates these patterns. Females oviposit on the leaves of plants in the cabbage family (Cruciferacea). Larvae feed on the leaves until they are ready to pupate. As anyone who has attempted to raise cabbages knows, high concentrations of larvae can rapidly defoliate plants, and can potentially eat all of their food source, resulting in starvation and death. Various studies have demonstrated that females can discriminate between different types of host plants, and prefer to oviposit on crucifers; specifically, those that are in prime condition, have no leaf damage, and show no evidence of prior use. Here we will study the oviposition site preferences of female butterflies by allowing them to lay eggs on different types of plants and cabbages that have been modified to simulate various physiological conditions and prior use by other females. We will then attempt to evaluate the importance of visual and olfactory cues to females seeking places to lay their eggs.

METHODS

Experiments will be conducted in a large plastic enclosure constructed by suspending transparent sheets from the ceiling to the floor of the lab. Exact size will be determined by the room available but the cage should be large enough to allow the butterflies free flight around the various oviposition sites being tested.

The flyway described in Exercise 4 on the perceptual abilities of bats is ideal for this analysis, although a smaller enclosure is fine. The butterflies themselves will be wild, caught prior to the analyses. Determine the sex of each butterfly (see figure) and release the males. Cabbages used in the study should be raised in the greenhouse, garden, or nursery, and need only a few leaves each (i.e., we do not need cabbage "heads," only a few leaves per plant).

Choice of host plant species

Place 10 plants each of 5 different species in the butterfly cage. Ten of the plants should be domestic cabbages, the remaining 40 should represent commonly occurring species collected from the same areas in which the butterflies were caught. Plants can be in individual pots or can be cuttings placed in jars of water. Arrange them 10 rows of 5 plants each. Ensure that all plants have approximately equal leaf areas. Release a female butterfly into the chamber. Record the amount of time she spends sitting on each type of plant. After 30 minutes examine each of the plants she visited for eggs and record the number of eggs on each. Remove any plant with eggs (save the eggs for later experiments) and replace it with a fresh plant. Repeat the experiment with another female, and continue repetitions until about 10 ovipositions have occurred.

Choice of individual plants

Use a large cork-borer to remove about half of the leaf area from each of five cabbage plants. Place these plants together with five unaltered cabbage plants in the butterfly cage. Introduce females one at a time and record their preferences as in the previous experiment.

Remove the defoliated plants and replace them with five cabbages that have eggs on them. These can be plants used in the preceding experiments or plants that have had eggs glued or smeared on their leaves. If eggs are not available, simply rub several leaves with the tip of a female's abdomen. Ensure that 5 unaltered plants are available in the cage. Introduce females one at a time and record their preferences as done previously.

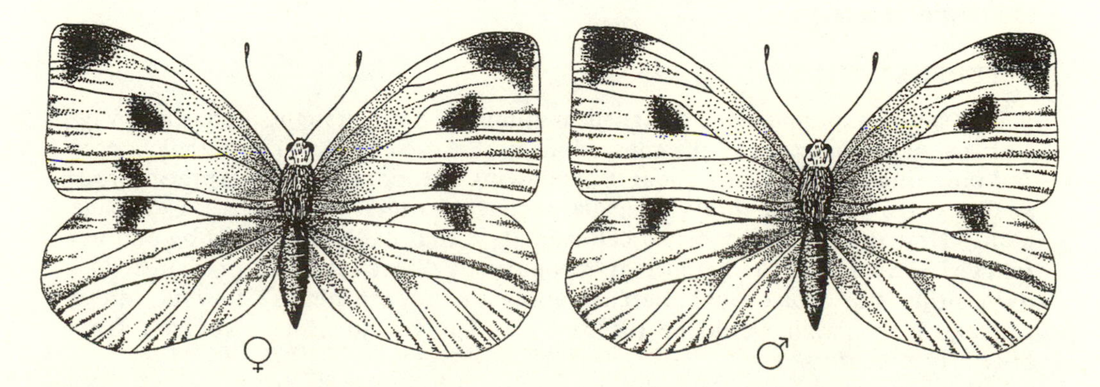

Female and male cabbage butterflies, Pieris rapae. *Note that the female has two spots on each forewing, whereas the male has only one.*

Cues used to identify sites

Prepare a cabbage extract by macerating several leaves in acetone. Apply this extract to 5 discs of filter paper, and treat 5 more discs with acetone as a control. Evaporate the acetone, place the discs in the cage, and allow females to choose among them as done previously.

Duplicate the experiment described above using cabbage extract-treated discs and extract-treated discs that have small glass beads arranged on them to resemble eggs.

ANALYSES

We are primarily interested in discovering whether the female butterflies can discriminate among various types of oviposition sites and choose those that are of the appropriate species and condition. We are secondarily interested in discovering the importance of olfactory and visual cues to any choice patterns.

Suggested null hypotheses and statistical tests include:

a. Females spent an equal amount of time on plants of all five species: Chi-square test, a priori expectation of equal frequencies. Alternatively: females spent equal average amounts of time on each plant species: Analysis of variance.
b. The number of ovipositions (or individual eggs) was equal for each species: Chi-square test, a priori expectation of equal frequencies. Alternatively: the average number of eggs (or ovipositions) was equal for all species: Analysis of variance.
c. Average time (or number of eggs) at whole and experimentally damaged plants were equal: Mann-Whitney U test.
d. Average time (or number of eggs) at clean and previously used plants were equal: Mann-Whitney U test.
e. Average time (or number of eggs) at cabbage-scented and unscented discs of paper were equal: Mann-Whitney U test.
f. Average time (or number of eggs) at clean and glass bead discs of filter paper were equal: Mann-Whitney U test.

Suggested graphical analyses include:

Bar graphs of the number of ovipositions, number of eggs, or time spent at each type of site.

INTERPRETATION

This set of experiments is designed to stimulate your thoughts about the ways animals make choices, and the consequences of these choices to both the chooser and the thing chosen. Think about the following questions when interpreting your analyses: How do you think a female cabbage butterfly recognizes a cabbage plant? How does she recognize a cabbage plant that has been visited by another female? Why is that an important recognition? What selective pressures does oviposition site choice by butterflies apply to cabbage plants? If you were a cabbage plant, what would be the best way for you to protect yourself from butterfly predators?

SUGGESTED REFERENCES

Mitchell, N. 1977. Differential host selection by *Pieris brassicae* (the large white butterfly) on *Brassica oleracea* (the wild cabbage). *Entomol. Exp. Appl.* 22: 208-219.

Rothchild, M. and L.M. Schoonhoven. 1977. Assessment of egg load by *Pieris brassicae* (Lepidoptera: Pieridae). *Nature* 266: 352-355.

Stamp, N.E. 1980. Egg deposition patterns in butterflies: Why do some species cluster their eggs rather than deposit them singly? *Am. Natur.* 115: 367-380.

Williams, K.S. and L.E. Gilbert. 1981. Insects as selective agents on plant vegetative morphology: Egg mimicry reduces egg laying by butterflies. *Science* 212: 467-469.

NOTES ON THE STUDY ORGANISMS

This analysis can be performed using many other species of butterflies, as long as the host plant of the species used is included in the choice experiments. A few cabbage plants grown outdoors will almost always attract cabbage butterflies or cabbage loopers (*Trichoplusiani*). Alternatively, adults are commonly found flying around private garden plots.

24. *Male Dominance and Mating Behavior in Blister Beetles*

ABSTRACT

Behavioral and morphological correlates of dominance are examined in male blister beetles. Effects of male body size on success in winning fights and on courtship activities are evaluated.

INTRODUCTION

Close examination of almost any behavioral pattern shows variation in the performance of the behavior by different members of a species. Such variation may be complicated, involving variable sequences of motor actions, or may be much simpler, involving variable duration or different intensity of actions. The degree of variation observed in a population presumably reflects the action of natural selection. In some cases, selection will reduce variability, as for example in the initiation of an escape response to the presence of danger. In other cases selection will increase variability. One situation in which we might expect such an increase is that in which the probability of a specific behavior benefiting an individual animal varies. If the animal can predict the probability of its behavior paying off, selection should favor such predictions, and the animal should vary its behavior patterns in response to its estimation of reward or success. For example, selection generally will not favor animals that all spend the same time and effort on aggressive interactions. When confronting a more formidable (perhaps larger) opponent, selection may favor the animal who retreats. When confronting a less formidable opponent, selection may favor an animal who escalates. Similarly, selection may affect courtship activities. If females are choosy about their mates, then males may evaluate their competition or their probability of successful courtship and modify their activities accordingly.

In this exercise, we will examine variability in the aggressive activities and courtship persistence of male blister beetles, *Epicauta pennsylvanica* (see figure). This species is chosen because of its abundance, and because it exhibits a relatively prolonged courtship, unlike most other beetles which have little or no courtship prior to copulation.

THE STUDY ORGANISM

Blister beetles comprise the family Meloidae, a diverse group that includes several important pests of potatoes, tomatoes, and other food plants. The life history of members of the genus *Epicauta* is complex, including what is referred to as hypermetamorphosis: larval stages each have quite different forms. The first larval instar is long legged and active. Depending on the species, it climbs a flower stalk and attaches itself to a bee or seeks out grasshopper eggs, which it eats. If it finds a bee, it rides back to the hive and feeds on bee eggs. Successive larval stages are more or less grub-like and may lack legs altogether.

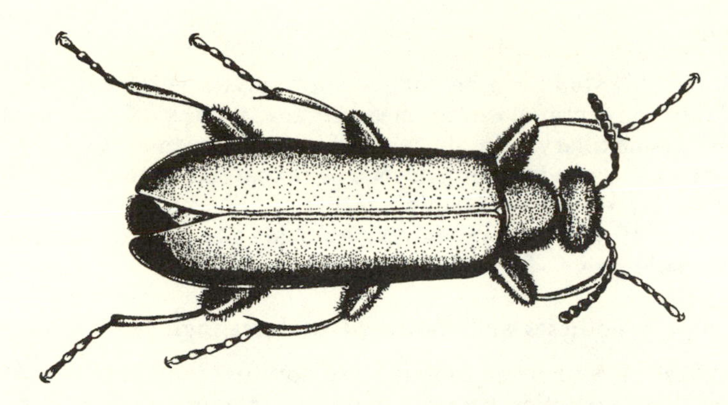

A typical blister beetle, Epicauta pennsylvanica.

Adults eclose after pupation and, in the case of *E. pennsylvanica*, feed on pollen, especially that of goldenrods and other late summer blossoms. Blister beetles receive their common name from the presence of canthardin, a substance that can raise blisters on skin or mucous membranes.

METHODS

We are interested in examining male aggressive interactions and courtship encounters.

Aggression

Locate a high-density population of blister beetles. Observe aggressive interactions from a distance of two m in order to avoid detection by the beetles. Aggressive encounters will generally take place when two or more males are in the presence of a single female. Define the winner of each outcome as that male who drives the loser from the presence of the female. After observing and recording each encounter, collect the interacting beetles and measure elytra lengths using a dial caliper. Place each beetle in a small vial, label the winner with a "W" and the loser with an "L" and rubber-band the two vials together. Repeat this procedure until a suitable number of observations have been made (approximately 50 per class). After returning to the lab, kill the beetles by freezing them, dry them in an oven (80°C for 2 days), and weigh the males if desired.

Courtship

To determine whether any size-dependent assortative mating is occurring, collect mating pairs of beetles. Measure elytra lengths for males and females, place each pair in a vial; dry and weigh after returning to the lab.

To determine the effects of male size on courtship, observe several courtship encounters. Record the duration of each encounter on a stopwatch. Collect the male, measure his elytra length, place him in a vial, and record his identity on the vial. After returning him to the lab, kill, dry, and weigh if desired.

ANALYSES

Analyses for this study can be conducted on elytra lengths or weights. Interested students may also wish to measure the widths of the basal antennal segments or the maxillary palpi, both of which have been suggested as secondary sexual characteristics under the influence of sexual selection (McLain, 1982). We are primarily interested in examining differences between average sizes of different classes of males and in regressing female size on male size and courtship duration on male size.

Suggested null hypotheses and statistical analyses include:

a. Median sizes of winners and losers were not different: Mann-Whitney *U* test.
b. There was no correlation between male and female sizes for beetles captured in copula: Pearson or Spearman correlation coefficient.
c. No linear relationship exists between the sizes of male and female beetles captured in copula: Linear regression.
d. No linear relationship exists between male size and courtship persistence: Linear regression.

Suggested graphical analyses include:

a. A scatter plot, with regression line, of the sizes of males and females captured in copula.
b. A scatter plot, with regression line, of the duration of male courtship as a function of male size.

INTERPRETATION

This set of experiments is designed to stimulate your thoughts about the relationships between body size, dominance and mating success. Think about the following questions when interpreting your analyses: Did size determine whether a beetle won an aggressive encounter? Did size determine whether a beetle mated? If larger beetles mate more often than smaller ones, why are there small beetles in the population? Did males who persisted in courting for longer periods of time mate more frequently? Why is there so much variability in courtship duration? Can you tell whether size or courtship duration is more important to mating success in this species? What does the fact that a male was captured while copulating tell about the male's fitness?

SUGGESTED REFERENCES

Erber, G.H. and N.S. Church. 1976. The reproductive cycle of male and female *Lytta nuttalli* (Coleoptera: Meloidae). *Canad. Entomol.* 108: 1125-2236.

Mason, L.G. 1972. Natural insect populations and assortative mating. *Amer. Midl. Natur.* 88: 151-157.

McCauley, D.E. and M.J. Wade. 1978. Female choice and the mating structure of a natural population of the soldier beetle *Chauliognathus pennsylvanicus*. *Evolution* 34: 174-180.

McLain, D.K. 1982. Behavioral and morphological correlates of male dominance and courtship persistence in the blister beetle *Epicauta pennsylvanica* (Coleoptera: Meloidae). *Amer. Midl. Natur.* 107: 396-403.

Selander, R.B. 1964. Sexual behavior in blister beetles (Coleoptera: Meloidae)I. The genus *Pyrota. Canad. Entomol.* 96: 1037-1082.

NOTES ON THE STUDY ORGANISMS

Old-fashioned potato beetles *(Epicauta vittata and E. pestifera)* are another meloid that can be substituted for *E. pennsylvanica.* These are pests of potato plants and may be common, especially in unsprayed, organic gardens. Locust borers (*Megacyllene robiniae*) also make a good experimental subject, and may be common in old field eco-systems in the late summer.

25. Mate Preferences of Male Guppies

ABSTRACT

The preferences that males show when given a choice of females are explored in experiments that contrast characteristics of females.

INTRODUCTION

Darwin (1871) suggested that many reproductive characteristics of males and females arise by means of sexual selection. He recognized two different but complementary aspects of sexual selection. First, selection may be the result of competition among individuals of one sex for access to the other sex. Alternatively, selection could take the form of choices made by one sex among individuals of the opposite sex. Since it is assumed that females take greater risks in reproduction, and because they reproduce less frequently than males, it is more likely for males to engage in intrasexual competition and for females to be the "choosier" sex (but see Gwynne, 1981). However, there are instances in which females congregate, and more females are receptive at any one time than a single male can inseminate. Hence, the opportunity for male choice sometimes exists.

We expect males to become choosey if females differ in quality, and males are confronted either with limited resources for reproduction or with the opportunity to choose among several females at any one time. Among species where egg number is a function of body size, there may commonly be opportunities for males to choose among females, since a male mating with a large female will father more young than one who spends the same time and energy mating with a smaller female. In many fish, fecundity is a function of size, and growth is indeterminate, suggesting that these opportunities occur frequently. Here we will explore some aspects of the choosiness of males guppies, *Poecilia reticulata*, when confronted by females of different sizes.

THE STUDY ORGANISMS

The general biology of the common guppy is described in exercise 8 on foraging patterns and prey choice. The critical details of guppy biology, as far as this experiment is concerned, are that guppy populations are frequently dense, that males compete for access to females, and that larger females produce more young than do smaller ones.

METHODS

We will examine male preferences for females of different types by confronting males with pairs of females. This is easily done using 40-1 (10 gallon) aquariums that have been filled with tap water and equipped with a heater and an aerator. Gravel and aquatic plants should not be used, since they obscure observation and are unnecessary. Do make sure that chlorine and fluorine have been removed from the aquarium before fish are introduced.

Since male guppies are stimulated to court more frequently when competitors are present (see experiment 27 on male-male competition in guppies), place five mature males in each of four 10-l aquariums. To evaluate effects of virginity on male choice, introduce two mature females that appear to be the same length, one of whom is a virgin and the other of whom is mated. Virginity should be ensured by raising the females in isolation from males. A mated female is clearly recognized by a darkened "brood spot" region of the abdomen (see figure). (Advanced pregnancy can be recognized by the eyes of the embryos, which are visible within this spot). After introducing the females, observe one of the males and record the total amount of time that he spends courting each of the two females on stopwatches. This is easily done if you use two watches, one for each female. (The actual courtship postures are described in the experiment on male-male competition). If enough observers and stopwatches are available, more than one of the males may be observed at the same time After 10 minutes have elapsed, remove the females, record the total times for each observation, and introduce a new pair of females. Repeat until at least 20 males have been observed.

To evaluate effects of body size, introduce two virgin females that differ in their total lengths by at least 10% of the larger one. Proceed as before, recording the total time spent courting each female by a focal male during a 10-minute interval. Repeat until at least 20 males have been observed.

ANALYSES

There are several ways in which we can analyze the data. We can score which female the male spent the most time with. Thus the number of minutes spent courting is reduced to counts of more time with the larger or more time with the smaller female. Under the null hypothesis that males do not discriminate, we would expect females to be courted equally, regardless of size or condition. That is, the number of males who spent more time with the larger female should be equal to the number of males who spent more time with the smaller female—a simple Chi-square test.

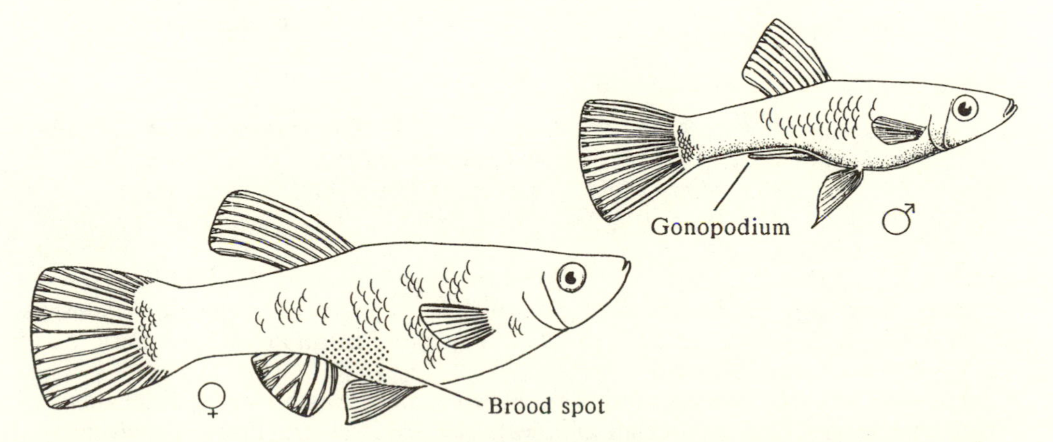

*Male and female guppies (*Poecilia reticulata*). The male is smaller, usually colorful, and has a modified anal fin (gonopodium) that is used as the copulatory organ. The brood spot is only seen on pregnant females.*

Alternatively, we may wish to use a *t* test to determine if there are differences in the proportion of time that a male spent with the larger or smaller female. To analyze proportions, it is necessary to perform an arcsin square root transformation on the proportions to achieve normality:

Arcsin square root transformations of selected percentages.*

%	ARCSIN	%	ARCSIN	%	ARCSIN	%	ARCSIN
0.0	0.0	1	5.7	10	18.4	25	30.0
0.2	2.6	2	8.1	12	20.3	30	33.2
0.4	3.6	4	11.5	14	22.0	35	36.3
0.6	4.4	6	14.1	16	23.6	40	39.2
0.8	5.1	8	16.4	18	25.1	45	42.1
				20	26.6	50	45.0

*To transform percentages greater than 50, subtract the appropriate arcsin$\sqrt{\ }$ from 90. For example: 80% = 100% − 20%

$$\begin{aligned} \arcsin\sqrt{80} &= 90 - \arcsin\sqrt{20} \\ &= 90 - 26.6 \\ &= 63.4 \end{aligned}$$

The advantage of this analysis is that it gives us an estimate of variability among males. This analysis may give us a different answer than the Chi-square test. If some males spend all of their time with the larger female and other males spend all of their time with the smaller female, then the variability among males may be too great to reveal an average difference.

INTERPRETATION

This set of experiments is designed to stimulate your thoughts about the ways that animals choose their mates, and, specifically, the criteria that males might use when choosing females. Think about the following questions when interpreting your analyses: Did males demonstrate any preferences? What does male preference for virginity or large size mean to the fitness of an individual male? Why do you suppose males don't simply court all females every time they encounter them?

SUGGESTED REFERENCES

Baerends, G.P., Brouwer, R. and Waterbolk. 1955. Ethological studies of *Lebistes reticulatus* (Peters). I. Analysis of male courtship patterns. *Behaviour* 8: 29-334.

Bateson, P. (ed). 1983. *Mate Choice.* Cambridge Univ. Press, Cambridge.

Darwin, C. 1871. *The Descent of Man and Selection in Relation to Sex.* John Murray, London.

Downhower, J.F. and L. Brown. 1981. The timing of reproduction and its behavioral consequences for mottled sculpins. In *Natural Selection and Social Behavior,* R.D. Alexander and D.W. Tinkle, (eds.). Chiron Press, New York, pp. 78-95.

Farr, J.A. 1980. The effects of sexual experience and female receptivity on courtship/rape decision in male guppies, *Poecilia reticulata* (Pisces: Poeciliidae). *Anim. Behav.,* 28: 1195-1201.

Gwynne, D.T. 1981. Sexual difference theory: Mormon crickets show role reversal in mate choice. *Science* 213: 779-780.

Loiselle, P.Y. 1982. Male spawning-partner preference in an arena-breeding teleost, *Cyprinidon macularius californiensis* Girard (Antherinomorpha: Cyprinodontidae). *Amer. Natur.* 120: 721-732.

Rowland, W.J. 1982. Mate choice by male sticklebacks, *Gasterosteus aculeatus*. *Anim. Behav.* 30: 1093-1098.

NOTES ON THE STUDY ORGANISMS

Guppies are easily raised in the lab and can be obtained from pet shops or tropical fish wholesalers. An 80-l (20-gallon) aquarium will support a large breeding population and yield a good supply of mated females for this study. Virgin females are harder to maintain because they must be raised in isolation from males. Do this by isolating young guppies in a 40-80 l tank. At least once weekly, remove all males, who can be easily recognized prior to full maturity by the developing gonopodium (see figure on page 101). In a juvenile male this fin will gradually lose its rounded shape and elongate. Alternatively, mature females may be isolated from males until the females have used all of their stored sperm. This may take 3-4 months. Because the females may produce several broods during this time, it will be necessary to remove all offspring to prevent them remating with their sons. If stock populations cannot be maintained, juvenile guppies can be obtained cheaply from tropical fish wholesalers who supply "feeder" guppies to pet shops where they are fed to piscivores. Any pet shop can order these young guppies.

26. *Female Choice and the Rare Male Effect*

ABSTRACT

Frequency-dependent selection as evidenced in a "rare male effect" is investigated using two strains of fruitflies *(Drosophila).*

INTRODUCTION

Sexual selection occurs when individuals of the same sex differ in their ability to obtain mates. These differences may be due to competition between members of one sex for mates of the other (intrasexual selection). Alternatively, differences may be due to mate preferences and choices made by members of one sex for mates (intersexual selection). Probably the most obvious intrasexual competitions occur between males for access to females: males of many species, from beetles to mice to lions, fight for mates. Mate choices are much harder to demonstrate and are frequently subtle. Both females and males of many species, including both vertebrates and invertebrates, do appear to be capable of discriminating among prospective mates, however.

In evolutionary terms, sexual selection is important because it can change the frequency of genes in a population. The actual type of change depends on the type of genetic traits that are selected. (Various types of change are discussed in exercise 21 on assortative mating in soldier beetles.) One very important change can be the reduction of genetic variability within a population. Suppose, for example, that females prefer a particular type of male as their mates. Males with the "wrong" genotype breed infrequently, if at all, and thus do not contribute to the gene pool of succeeding generations. Males with the "right" (preferred) genotype pass on their traits, but the variability within the entire population is reduced. If selection were strong, variability might be eliminated altogether. There are many ways in which variability might be preserved, and one of them actually involves female choice for males having "right" or desirable traits. This type of choice is not based on absolute criteria (i.e., there is no absolutely biggest, brightest, showiest, etc.), but on *relative* differences among males. Specifically, the attractiveness of a particular phenotype is dependent on its rarity. The rare male is the one most preferred. As his genes become common, they lose their attractiveness. Thus the same phenotype can be wildly successful or dismally unsuccessful, depending on its frequency.

Rare male effects have been reported for a variety of species, ranging from fruitflies to fish. The phenomenon, however, has not been found by all investigators, and its mechanism, and even its existence, remain the subject of much controversy. Here we will examine rare male mating advantage under controlled conditions.

METHODS

Obtain virgin female and male fruitflies (*Drosophila* spp.; see figure) that have been separated and isolated for four days following emergence. Use two strains

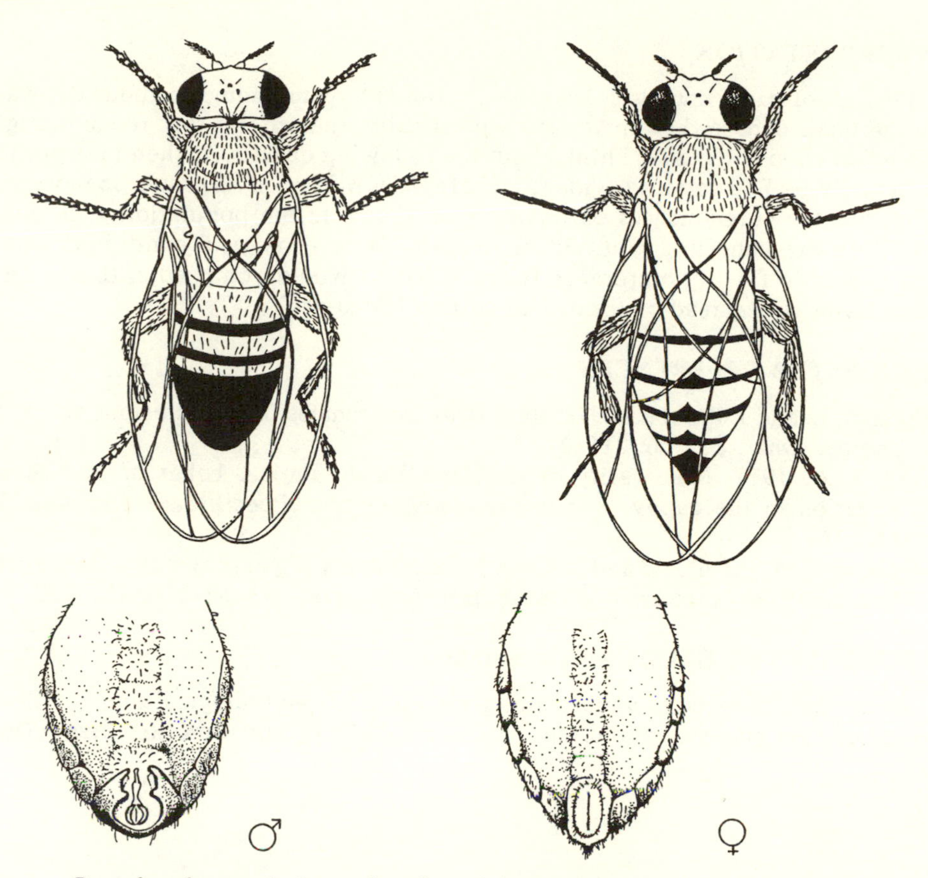

Dorsal and ventral views of male and female Drosophila melanogaster.

that differ visibly in their phenotypes. Commonly available traits include "ebony" (a dark body color), "curly wing", and "white eye." Reciprocal tests will hopefully reveal any strain differences in vigor or activity levels.

Each test should be conducted in a standard fruitfly culture vial. Place eight males of type "a," two males of type "b," and a female in a vial and observe until the female has copulated with one of the males. Repeat using a female of the other strain.

Repeat the entire experiment reversing the relative abundance of males (i.e., eight males of type "b" and two of type "a"), again replicating with females from each strain.

Continue replications until between 25-50 females of each type have been observed.

ANALYSES

We are primarily interested in knowing whether male mating frequency is proportional to male abundance. If females were indiscriminate, those males making up 20% of the population should account for 20% of the copulations. If females show preferences based on rarity, this will not be true. If females show preferences based on strain, then one strain should always "win" regardless of its relative abundance. Use the Chi-square test, *a priori* expectation, to evaluate these hypotheses.

INTERPRETATION

This set of experiments is designed to stimulate your thoughts about the ways that animals choose their mates and, specifically, the criteria that females might use when choosing males. Think about the following questions when interpreting your analyses: Did you find evidence that females preferred rare male phenotypes? What are the consequences of this type of choice to the population gene pool? What are the benefits to individual females? What ecological conditions would you expect to favor a preference for rare males? Would you predict this preference to be widespread or limited to certain life styles?

SUGGESTED REFERENCES

Ehrman, L. and J. Probber. 1978. Rare *Drosophila* males: The mysterious matter of choice. *Amer. Sci.* 66: 216-222.

Farr, J. A. 1977. Male rarity or novelty, female choice, behavior and sexual selection in the guppy, *Poecilia reticulat*a (Pisces: Poeciliidae). *Evolution* 31: 162-168.

Markow, T. A., M. Quaid and S. Kerr. 1978. Male mating experience and competitive courtship success in *Drosophila melanogaster*. *Nature* 276: 821-822.

NOTES ON THE STUDY ORGANISMS

Drosophila of many mutant strains are available from biological supply houses, as are culture materials and directions for maintaining and obtaining virgin flies.

27. *Male-Male Competition Among Guppies*

ABSTRACT

Competition for mates among male guppies is examined in populations having different sex ratios.

INTRODUCTION

Competition among males for mates is a common and potentially forceful form of sexual selection. The reproductive rewards to males who successfully outcompete their rivals may be very great, especially when a single male can monopolize many females. Such monopolization sometimes occurs because females are gregarious and clumps of females can be defended by single males (e.g., some African antelopes). Alternatively, monopolization may occur when males successfully exclude competitors from superior resources (e.g., red-winged blackbirds), necessary breeding sites (e.g., dragonflies, elephant seals), or socially desirable mating spots, as on leks (e.g., sage grouse). Workers dating back to Darwin have argued that many of the sexually dimorphic weapons possessed by males but not females can be attributed to the intrasexual selection resulting from male competition for mates. Thus male antlers, horns, canine teeth, etc., are thought to have evolved for use by males against competitors in the struggle for mates.

Ecologists frequently distinguish between interference and exploitation during competitive interactions. **Interference competition** involves one individual inhibiting another individual's access to some resource. **Exploitation competition** involves one individual reducing the amount of a resource available to others without interfering with their access. The monopolization of territories, resources, females, and socially desirable mating spots described above are all examples of interference competition, since the presence of the monopolizer excludes competitors. Some animals do not monopolize resources, and compete through exploitation. Male poeciliid fishes, for example, do not provide resources to females, and cannot directly monopolize mates. These males increase their chances of successfully mating by increasing the rate at which they court females. More attempted matings usually means more actual matings. This is due to the facts that males appear to be able to inseminate females without the female's consent, and that females of some species appear to prefer to mate with males having higher courtship rates. Males courting at high rates pay a price, however, since predators selectively eat the brightly-colored, flashy, courting males. In this exercise, we will evaluate exploitation competition among male guppies (*Poecilia reticulata*) in populations having different numbers of competitors.

THE STUDY ORGANISMS

The general biology of the common guppy is described in exercise 8 on foraging patterns and prey choice.

METHODS

Establish populations of guppies in 4-l goldfish bowls 15 minutes prior to testing. Each population should consist of ten fish, but the sex ratio should vary. If enough bowls are available, populations having 9:1, 5:5, and 1:9 male:female ratios should be established in replicate or triplicate. Mask all but one side of each tank with opaque paper to reduce disturbance of the fish during observation.

Prepare for data collection by reviewing the motor patterns involved in guppy courtship and mating. A detailed account is provided by Clark and Aronson (1951). An abbreviated account follows:

1. *Gonopodial swinging*. The male swings his modified anal fin (the gonopodium) forward so that it projects anteriorly. Swings alternate from left to right sides. The dorsal fin is held erect at the same time that the anal fin is swung.
2. *Body curving*. The male holds his body in either an arc or an "s" shape, usually perpendicular to the female.
3. *Thrusting*. The male swings his gonopodium forward as he swims along side the female. He may or may not actually touch the female's body.
4. *Copulations*. These may be of variable duration, but are always fairly quick-less than three seconds. They may follow thrusting, and are distinguished by the fact that the female remains stationary during genital contact.

Observe each male in each tank for 15 minutes. Count the number of each courtship pattern given, and record the total duration of courtship behaviors on a stopwatch.

If the males do not court, remove the females from each population and introduce an equivalent number of new (better yet, virgin) females.

ANALYSES

Because courtship in guppies is costly in terms of exposure to predators, we expect males to court only when necessary. At the same time, courtship is necessary for mating, and the frequency of mating is a function of courtship rates when more than one male is present in a population. We are therefore interested in knowing whether courtship rates increase as the number of competitors increases.

Suggested null hypotheses and statistical tests include:

a. The mean numbers of each courtship activity (or of all activities related to courtship) were equal for all sex ratios: Analysis of variance.
b. The mean time spent courting was equal for all sex ratios: ANOVA.

Suggested graphical analyses include:

a. Bar graphs for each courtship activity as a function of sex ratio.
b. Bar graphs of time spent courting as a function of sex ratio.

INTERPRETATION

This set of experiments is designed to stimulate your thoughts about intrasexual competition for mates. Think about the following questions when interpreting your analyses: Did male guppies increase their courtship rates when competitors were present? Why didn't all of the males court females? Did you find any indication that males tended to do whatever the other males were doing (i.e., if the others were not courting, no one was; if the others started to court, everyone did)? Why do you suppose this might be selectively advantageous to individual males? Why do you suppose this might be selectively advantageous to individual females? Did females incite the males?

SUGGESTED REFERENCES

Clark, E. and L.R. Aronson. 1951. Sexual behavior in the guppy, *Lebistes reticulatus* (Peters). *Zoologica* 36: 49-66.

Farr, J.A. 1974. The role of predation in the evolution of social behavior of natural populations of the guppy, *Poecilia reticulata* (Pisces: Poeciliidae). *Evolution* 29: 151-158.

Farr, J.A. 1976. Social facilitation of male sexual behavior, intrasexual competition, and sexual selection in the guppy, *Poecilia reticulata* (Pisces: Poeciliidae). *Evolution* 30: 707-717.

Farr, J.A. and W.F. Herrnkind. 1974. A quantitative analysis of social interaction of the guppy, *Poecilia reticulata* (Pisces: Poeciliidae) as a function of population density. *Anim. Behav.* 22: 582-591.

Gandolfi, Gilberto. 1971. Sexual selection in relation to social status of males in *Poecilia reticulata* (Teleostei: Poeciliidae). *Boll. Zool.* 38: 35-48.

NOTES ON THE STUDY ORGANISMS

Maintenance of guppies is discussed in exercise 25 on mate preferences of male guppies. This experiment works best with virgin females.

An Introduction to the Analysis of Data

In each of the exercises we pose a series of questions that we wish to answer. In order to answer them we collect sets of data that are relevant to the questions that we asked. We do not collect data indiscriminately, but rather use the questions to identify the kinds of data that we need to collect. The kinds of questions that we pose determine the kinds of data we collect, and the way in which we analyze the data. Science works this way: it poses questions, collects data, and analyzes the data to answer the questions. We must add that often the answer to a question generates new questions, new experiments, and a new round of data collection and analysis.

In this introduction we can only give you a flavor of methods in data analysis. We emphasize analysis and quantification because they lead to greater insight and clearer communication. For example, a common phrase that plagues us is, "You know what I mean." Actually, we often don't; yet our acceptance of that phraseology implies understanding. In the development of an experiment we attempt to state what we mean as clearly as possible. How we frame a question determines how we go about finding the answer (or even if it is possible to find an answer). We collect data that bear on the question, we analyze these data, and we interpret the results of the analysis. We may disagree on the interpretation of the results, but if we are careful, we will understand the basis for our disagreement, and that will lead to a fuller understanding of questions and processes.

OBSERVATIONS, SAMPLES, AND POPULATIONS

All scientific inquiries are based on observations. These are simply measurements of the variables we have chosen to study; that is, they are *quantities* that indicate the various states each variable had in our study. If, for example, we were interested in the variable "weight," then the observations we made were the weights of the individuals we studied.

Our sample is the set of observations that we collected. The sample size is the number of observations in that set. The statistical population is the totality of all of the possible observations of the variable in question, whether they were ever made or not. For example, if we weighed 50 male Japanese beetles, we would have collected a sample of 50 observations from the statistical population of the weights of all male Japanese beetles in the study site. While it may sometimes be possible to actually measure all individuals in a population, in all of the exercises presented here, we measure a sample that is much smaller than the entire population. This is true simply because the statistical populations we study are very large, and sampling is time consuming, and may be tedious and expensive.

KINDS OF QUESTIONS

We begin any inquiry with a question. This is nothing more than formalizing our curiosity. Some people seem more curious than others, but that is probably a reflection of having asked more questions, or being asked to ask questions. Like most things, we get better with practice, and we often have had little practice in asking questions.

There are basically two kinds of questions that we will ask. The first is **descriptive**. Descriptive questions include: What is it? How big is it? How many were there? How variable were they? Descriptive questions concern our sample. For example, we might ask how many kinds of fish we collected from a particular pool. Or we might ask what was the length of the longest fish that we caught. (Note that in the question we specified length. We could have said "how big," but "big" is ambiguous, since eels are often long, but we all might not agree that they are big).

In answering descriptive questions, we use **statistics**, which we will introduce a little later. We will give preference to those statistics which allow us to make **inferences** about the populations we have sampled.

Inferential questions make up a second class of questions. Inferential questions are comparative. They include: Is this group of fish longer than that one? Are men taller than women? Is this population more variable in a specific character than another population?

Suppose, for example, that we have measured the lengths of the elytra (wing cover) of male and female soldier beetles (*Chauliognathus pennsylvanicus*). A typical data set for such measures of size might look like this:

Table A. Elytra lengths (mm) for male and female soldier beetles, *C. pennsylvanicus*.

MALE	FEMALE
7.85	7.90
8.01	7.88
7.56	7.95
7.40	8.15
8.21	7.43
8.00	7.24
7.15	8.15
7.23	8.00
7.63	7.81
7.71	7.89

We might be interested in describing the soldier beetle sample. For instance, males ranged from 7.15 to 8.21 mm, females from 7.24 to 8.15 mm. At the same time, we might be interested in inferring whether males and females were the same size on average. If all of one sex were larger than the other, we could answer this question by inspection of the data set. As is generally the case, however, there is much overlap in sizes, and this overlap requires that the data be **analyzed** to reveal an answer to the question of comparative sizes. We will deal with appropriate analyses in the following text. For now, you should be aware of the difference between descriptive questions (How long? How variable? What was the average? What was most common?) and inferential questions (Is this longer than that? Were these more variable than those? Were the averages different from each other?).

KINDS OF DATA

Science deals with things that we can count or measure. These measurements are the **data** (plural of datum) that we analyze to answer questions.

Two kinds of data concern us. The first are things that we count. In the table below we have tabulated the number of errors made by the big brown bat (*Eptesicus fuscus*) when it was confronted with obstacles of different sizes.

Table B. Performance of *Eptesicus fuscus* when confronted with obstacles of two sizes.

	OBSTACLE SIZE LARGE	SMALL
Number of collisions	25	40
Number of misses	200	170
Total passes	225	210

In this case we recorded what happened when the bat approached obstacle. Either it hit it or it missed it. There is no in between; it hit it softly, it missed it by a mile. The hit-and-miss categories are **discrete.** Sometimes we say that these kinds of outcomes are "known without error."

Not all measurements can be known without error. The soldier beetle data cited earlier (Table A) provides an example. The first male had an elytra length measured at 7.85 mm. Was this the exact length? It may have actually been 7.850, or 7.851, or something else. The **precision** of the measuring calipers only allowed measurements of hundredths of millimeters. More precise instruments might allow finer measurements to be made, but no degree of precision would give the exact measurement of the beetle's wing covers. Such measurements are called **continuous variables.** No matter how finely we divide the measurement scale, we will not produce precise discrete classes. Lengths, weights, volumes, time intervals, and compass bearings are all examples of continuous variables that we will encounter.

How we measure continuous variables may define the answers we find to the questions we ask. If we measured soldier beetles to the nearest centimeter, we would find that they all had a length of 1. Obviously, this would obscure any differences between sexes or populations. On the other hand, if we measured them to the nearest nanometer, we would spend much time, effort, and money, and probably not gain any more information than if we had used dial calipers accurate to hundredths of millimeters. Since our measurement technique will cause us to view a truly continuous variable as discrete classes, we must choose units appropriate to the question. Sometimes units will be dictated by available equipment or experimental design. If a surveyor's transit is not available, a hand-held compass may do. Other times, some understanding of the biological system under investigation will help choose appropriate units. Knowledge that soldier beetles have elytra lengths varying by one mm or less allows choice of tenths or hundredths of a mm as suitable units.

The soldier beetle data in Table A is simply a list of lengths recorded as they were measured. While it may sometimes be convenient simply to write down each observation in a list, it is usually necessary to summarize the observations for analysis. One common summary technique involves listing the frequencies of different measurements, and this may require that the scale be changed after the data were collected. If, for example, we were to record the frequencies of measurements for male beetles to the nearest hundredth mm, we would simply restate the data in Table A. If we changed the scale to the nearest two-tenths mm, we would reformulate the data as shown below.

Table C. Frequencies for elytra lengths for soldier beetles listed in Table A.

LENGTH (MM)	NUMBER OF MALES	NUMBER OF FEMALES
7.00-7.19	1	0
7.20-7.39	1	1
7.40-7.59	2	1
7.60-7.79	2	0
7.80-7.99	1	5
8.00-8.19	2	3
8.20-8.39	1	0

The type of summary shown in Table C will commonly make analyses easier, and make patterns clearer. For example, inspection of Table A does not show that females were most commonly between 7.8 and 8.0 mm, or that males were more variable in length. These trends are much clearer in Table C. Such summaries will be very important in many of the statistical analyses that follow, and in some of the graphical analyses described later.

KINDS OF VARIABLES

When designing an experiment we often want to examine the responses of individuals to a particular treatment. In the case of the bat data we have established two kinds of obstacles: large and small. These kinds of variables are called **independent.** They are not influenced by the response of the bat; the size of the obstacle did not change whether it was hit by a bat or not. On the other hand, whether or not a bat hit an obstacle could be **dependent** on the size of obstacle. Often we establish an independent variable, or treatment, when we conduct an experiment.

The distinction between independent and dependent variables is often not clear, and we may measure two or more dependent variables. For example, we might measure the length and weight for a sample of beetles. Neither is really independent of the other. If, on the other hand, we also recorded sex, then we would list sex as an independent variable (weight does not change sex in beetles) and recognize that both length and weight are dependent variables.

VISUAL SUMMARIES

After we have collected a set of data we want to analyze it and present the results of that analysis. Often we will want to summarize the data and its analysis in a series of figures and tables. We will do this because it is important for the reader to have access to the relevant information upon which you based your conclusions.

Figures and tables share one element in common: they must be able to stand alone. That is, all of the information required to understand a figure or table must be included in it. We will examine each in turn.

Figures

A figure is composed of two elements: the figure itself and its associated legend.

The **legend** of a figure describes what the figure is about. In addition, the legend includes any detail that will facilitate interpretation of the figure. Several examples are given on the next pages, but you can also study the figures in any of the journals in the library.

The construction of a figure is an aesthetic exercise. It requires that you be clear in regard to the purpose of the figure. Consider a simple plot of height versus age. This is a plot of points on a set of X and Y coordinates. The X axis, or abscissa, runs horizontally, while the Y axis, or ordinate, runs up and down. First we have to decide which measure, height or age, is on which axis. By convention, the X axis is used for the independent variable and the Y axis is used for the dependent variable. In our example, age would go on the X axis.

If we are dealing with two dependent variables, then we must choose which one is on which axis. For example, if we have weights of male and female soldier beetles that have been captured in copula, either sex might appear on the X axis. In all figures each axis is labeled and the units of measurement are specified. When a reader examines a figure in your paper, the variables, the scale, and the range of measurement should be clear (Figure 1).

When we construct a figure we are trying to illustrate a pattern. If this is a plot of height and weight for members of the class then those points should be large enough not to be mistaken for fly specks, and small enough to not exaggerate the values being plotted (Figure 2).

Thus far we have described a figure that plots a set of points. We should mention two other kinds of plots: bar graphs (Figure 3), and frequency histograms (Figure 4). Bar graphs are used to illustrate discrete data. In them we cannot link any two categories together, hence we use separate bars to illustrate the observed values in any class. By contrast, a frequency histogram illustrates the number of observations (frequency) in each class in a continuous distribution.

For illustrative purposes, a figure with 6-10 classes "looks right" (Figure 4). If there are more than 10 classes, your figure will look cluttered; if there are fewer than 6, no patterns will emerge. But remember, you must control the figure so it presents the data as clearly and as accurately as possible.

Tables

The rules for constructing a table are much the same as those for a figure. The important point is clarity. We often choose a table rather than a figure because the data set is too small to warrant a figure (recall the bat data in Table B) or it is large and complex and not easily illustrated. Many publishers prefer tables to figures because they are cheaper to publish.

It is possible to include a great deal of detail in a table, but remember both tables and figures are means to summarize your data. Tabulating large sets of raw data does not further your arguments.

Tables also have labels or titles, called captions. Captions are always placed at the top of the table.

STATISTICAL SUMMARIES

When we want to discuss our findings and draw conclusions regarding them, we need more rigorous methods than graphs or tables alone. Various kinds of statistical analyses provide the means to accurately describe the results of our

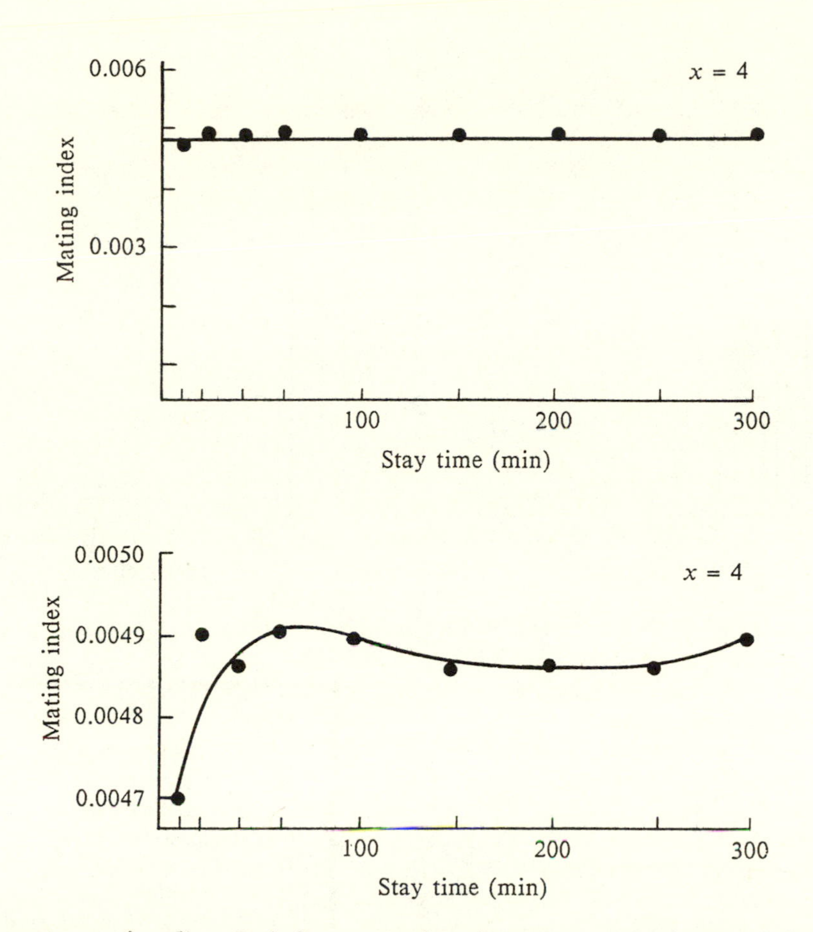

The importance of scaling. Both figures graph male mating success (mating index) for an insect (Scatophaga) *as a function of how long the male and female remain mated (stay time). The points in each figure are identical, but the scaling has changed, as has the interpretation. (From Curtsinger, J.W. 1986. Stay times in* Scatophaga *and the theory of evolutionarily stable strategies.* Amer. Natur. *128: 130–136.)*

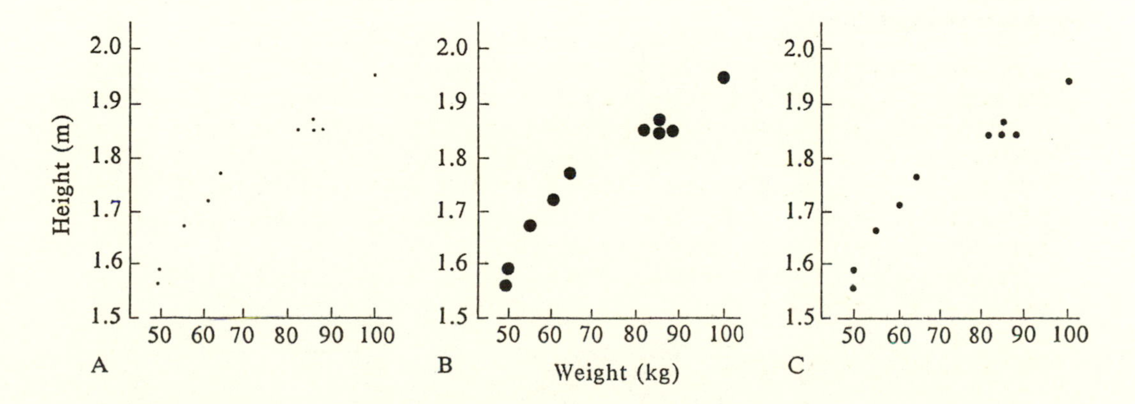

*Examples of incorrect (***A** *and* **B***) and correct (***C***) point sizes in scatter plots of height vs. weight for humans.*

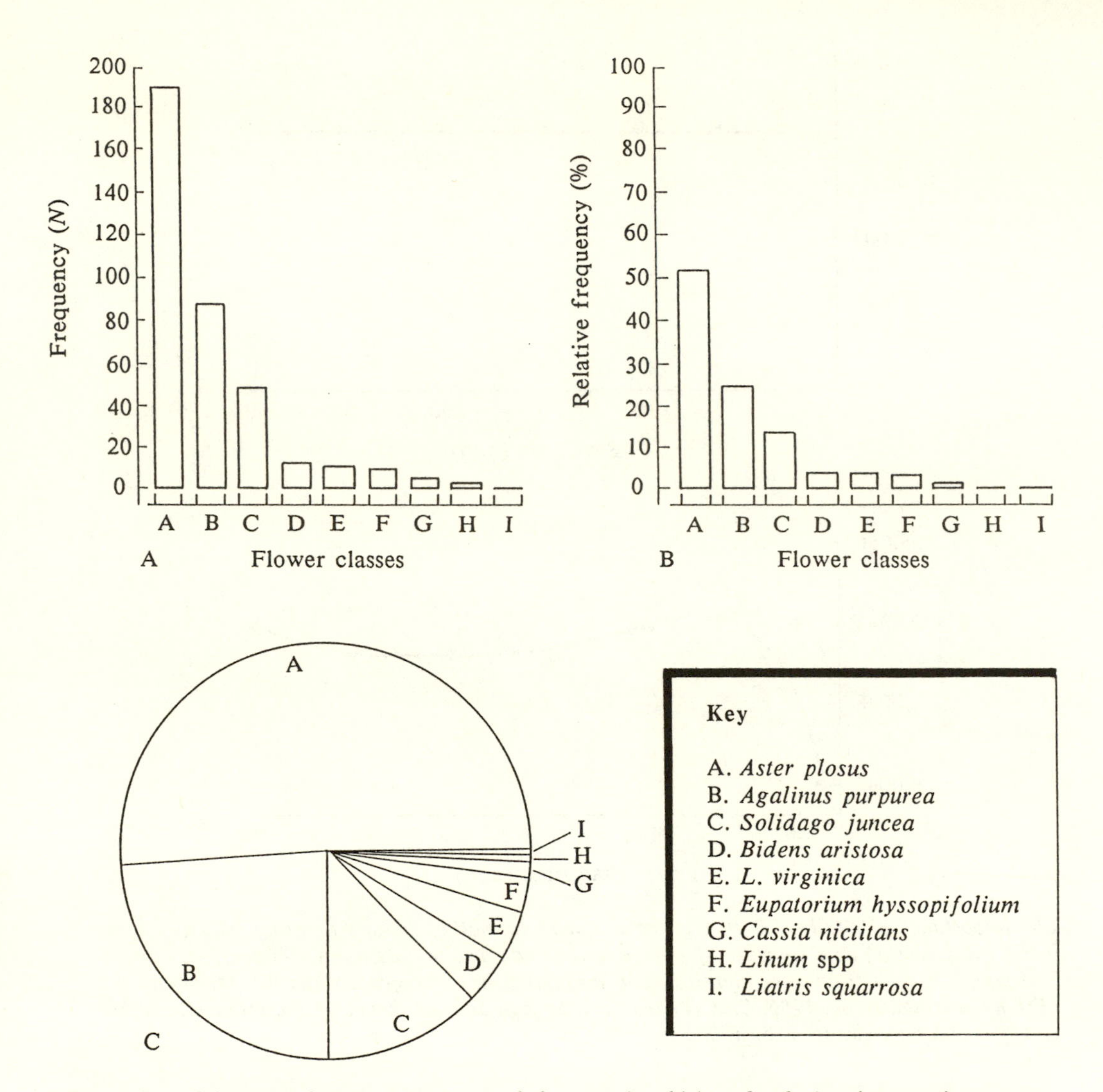

*Examples of bar graphs presenting actual frequencies (***A***) and relative frequencies or percentages (***B***). The pie diagram (***C***) presents the same data on flower abundance that is given in the other two illustrations.*

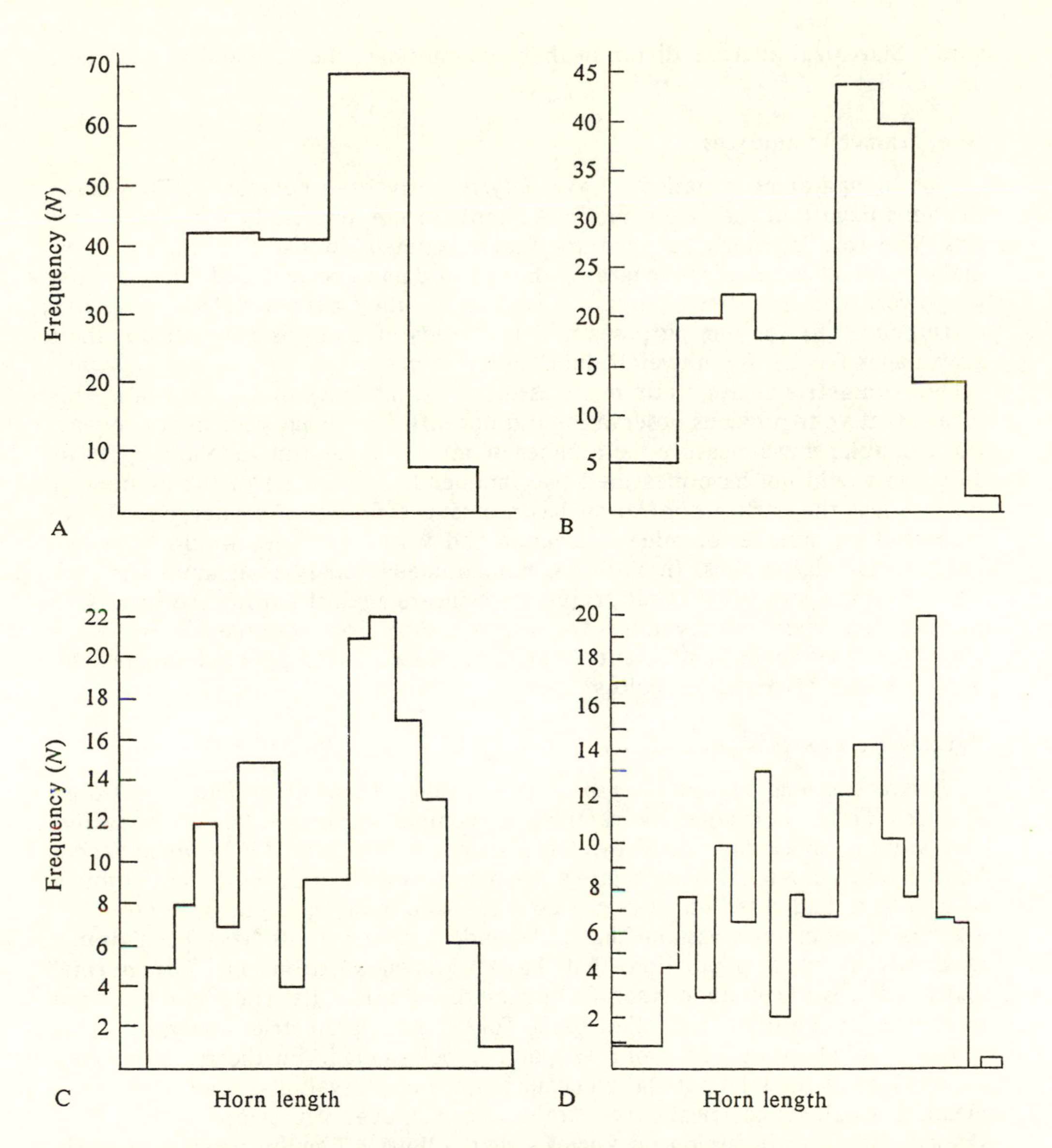

*Frequency histograms having too few (**A** and **B**) and too many (**D**) classes. Histogram **C** highlights the bimodality of the distribution. All of these histograms present the same data on length of the thoracic horn in male forked fungus beetles,* Bolitotherus cornutus.

study. Statistical analyses differ in their assumptions, their diversity, and their ease of use.

Nonparametric analyses

Most nonparametric statistics have only recently been developed. They have the advantage that they are sometimes simple to use, especially with small data sets. The way in which the data are analyzed makes intuitive sense, and they make minimal assumptions regarding the sets of data to be analyzed. They are not as powerful as parametric procedures—that is, they cannot detect as small a difference. But for our purposes and the kinds of analysis we will do, their advantages frequently outweigh their shortcomings.

Nonparametric analyses minimally assume independence of observations. This means that your previous observation did not influence what your next one was. For example, if we measure the number of mice in a field in June and again in July, we would not have measured two independent estimates of the number of mice, since the earlier population has a strong effect of the later one. If we measured the number of mice in London and Washington, we would have two independent data points. In addition, nonparametric analyses assume that each observation is selected without prejudice to insure against biasing the results by picking the "best" observations for the analysis. For example, if we were studying mouse colors in different populations, we would sample randomly to avoid overestimates of particular colors.

Parametric analyses

Parametric analyses are the most fully developed and historically most used analyses. They will handle large samples and complex experimental designs easily, and are the most commonly available forms of analysis. Many inexpensive hand-held calculators will calculate a mean and a standard deviation at the touch of a button, and these statistics are used to make many inferences.

Parametric analyses assume that independent observations have been drawn randomly from a population that has a known distribution. The **normal distribution** is a well understood frequency distribution that is common but not universal in nature. It is the basis for many parametric analyses. It is symmetrical about its center, or mean, and is "bell shaped." Furthermore, defined proportions of the total population occur at known intervals above and below the mean. Thus all symmetrical curves are not normal, even if they are roughly "bell shaped" (see the discussion on kurtosis that follows). Finally, most parametric inferences assume that the samples being compared have equal variances, the variance being a measure of the spread of the distribution (i.e., the width of the bell).

Circular statistics

Suppose we release a homing pigeon and record the direction in which it vanishes: 10° to the east of due north. Now we release a second pigeon and follow it. It vanishes at 5° to the west of due north, or 355°. The third bird we release flies due south (180°). If we compare the differences among the three birds on a linear scale we conclude that birds 1 and 2 were more like bird 3 (10 - 180 = -170; and 180 - 355 = -175) than like each other (10 - 355 = -345). Yet it is obvious that this conclusion is wrong. Our erroneous conclusion is a consequence of dealing with what is termed a **circular distribution.**

Both compass bearing and time measures (either daily or seasonal) are circular. Because these distributions fold back on themselves, they require special handling. They do not make any new assumptions, and both parametric and nonparametric analyses are available for circular measurements.

DESCRIPTIVE STATISTICS

We can describe any distribution of numerical observations by estimating the center of the distribution and the scatter of observations around that center. The center of a distribution is estimated from the sample using one or more measures of central tendency, sometimes called measures of location. The scatter of points around the center is evaluated using a measure of dispersion or variability.

Measures of central tendency

These measures can be used for both discrete and continuous data. They estimate the center of a distribution and the location of that center on a real number line: $-\infty < \text{center} < +\infty$.

a. **Mode.** The most frequent class. The mode is simply the most common value or class in the distribution.

b. **Median.** If all of the observations are arranged in rank order from smallest to largest, the median is that value which is bounded by 50% of the observations on each side. If there is an odd number of observations in the distribution, then the median is simply the middle value in the ranked observations. If there is an even number of observations, then the median is equal to the sum of the two observations closest to the middle divided by 2.

c. **Mean.** The arithmetic average of all observations in a distribution. The mean is equal to the sum of all observations divided by the sample size.

$$\text{Mean} = \overline{Y} = (\Sigma Y_i)/N$$

where Σ means "the sum of." Y_i is the ith observation, where i can have values from 1 to N; and N is the sample size or the number of observations. In general, a variable with a bar ($\overline{Y}$) over it is a sample mean.

If the observations are drawn from a normal distribution, then the mean = the median = the mode, because the normal distribution is symmetrical about its center.

Measures of variability

Individuals in a sample differ from one another. In order to describe those differences we need some way to summarize variation around the center of the distribution.

a. **Range.** The range of a distribution is the difference between the largest and the smallest value.

$$\text{Range} = \text{Largest } Y - \text{Smallest } Y$$

b. **Variance** and **standard deviation** of a sample. The variance is a measure of the average squared deviation from the mean. The standard deviation is the square root of the variance.

$$\text{Variance} = s^2 = \Sigma(Y_i - \overline{Y})^2/(N - 1)$$

$$\text{Standard deviation} = s = \sqrt{s^2}$$

It is inefficient and time-consuming to calculate a variance using the above formula, because subtracting each observation from the mean and squaring it adds to errors in the calculation.

If we restrict our attention to the numerator which we will call y^2 (n.b., this "y" is lower case), then the following formula allows direct calculation of the variance.

$$\text{Variance} = \Sigma y^2/(N - 1)$$

and

$$\Sigma y^2 = \Sigma Y_i^2 - [(\Sigma Y_i)^2/N]$$

In using this formula we need only know the sum of the observations (the sum of Y), the sum of each observation squared (the sum of Y^2) and the sample size. Note that the sum of Y is also used to calculate the mean. Since this formula for calculating the variance (and standard deviation) uses our original observations, the rounding errors inherent in the previous formula are avoided.

HYPOTHESIS TESTING

Hypothesis testing is a formal way of asking and answering questions. In its simplest terms it involves stating an hypothesis and either rejecting it or not on the basis of some criteria.

The null hypothesis

All hypothesis testing involves discrimination between pairs of alternative hypotheses. The simpler member of this pair is called the **null hypothesis.** The null hypothesis specifies or implies a distribution and its parameters, and thus enables us to calculate expected values to compare with sample values. It is, as the name implies, the hypothesis of no effect or no difference. The other hypothesis in the pair covers all possible alternatives to the null hypothesis, and is thus called the **alternative hypothesis.** Only one of the two can be true, and the simpler null hypothesis will be accepted unless there is good reason to reject it.

Suppose, for example, that we wished to compare the lengths of the male and female soldier beetles that we introduced earlier. Our null hypothesis might be that the two length distributions were drawn from the same common distribution (i.e., that there is no difference). Our alternative hypothesis is that males and females have different length distributions. We proceed to test the null hypothesis with an appropriate statistic. If we reject the null hypothesis, we are left with the alternative that there is a difference. There is no other option.

The set of hypotheses that we choose will be determined by the questions that we ask. In the soldier beetle example, we asked whether males and females were of different lengths. Rejection of the null hypothesis means that we have data that shows a difference; it does *not* mean that males are smaller than females. It means that males are either larger or smaller than females, and we don't know which. This analysis involved what we call a two-sided alternative. Inequality means that one is bigger than or one is smaller than the other. We could have begun with a different set of one-sided hypotheses that would address the question of which is larger.

Suppose we want to know whether females are the larger sex, possibly because this is the case in related species, or because past studies have shown that females are bigger. Regardless of the reason, we now want our null hypothesis to state that females are smaller than or equal to males, and our alternative to state that females are bigger than males. The null hypothesis still includes the possibility of no difference (equality) and is still the simpler argument. Rejection of the null hypothesis leaves the alternative that females are larger.

There is, of course, another set of one-sided hypotheses. Should we have reason to suspect that females are smaller than males, our null would be that they are greater than or equal to males, and the alternative would be that they are smaller than males.

The hypotheses that we evaluate are simply formal phrasings of the questions we ask. As in conversation, we must pose the question before we answer it. The set of hypotheses to be tested must be chosen before, not after, the analysis and decision making. Blindly applying statistical analyses to a set of data in the hope of finding some significant answer is an abuse of statistics. The decision to test two-sided or one-sided hypotheses must be made before proceeding with the analysis.

Criteria for decisions

When we introduced the idea of a normal distribution we indicated that it was well understood. This means that the shape of the distribution is precisely known. Suppose, for example, that female soldier beetles in northern Virginia have an average elytra length of 8.00 mm, and that the standard deviation for length is 1.00 mm. If female length has a normal distribution, 68% of all females will have lengths within ±1 standard deviation of the mean (i.e., 7.00-9.00 mm); 95% will have lengths within ±2 standard deviations (6.00-10.00 mm); and 99.8% will have lengths within ±3 standard deviations (5.00-11.00 mm).

Now, suppose someone found a female that was 12 mm long. Is it likely that she is from the original population? It doesn't seem likely, does it? She is too long—much longer, in fact, than 99.8% of the individuals in the total population.

All statistical inference is based on distributions that indicate the probability of certain events occurring. We used a normal distribution to determine the likelihood that the very large female was drawn from our sample population. We will use variations on this approach to compare populations and to generate our own distribution for events.

Significance level

Suppose someone found a female who was 11.0 mm long. Is it likely that she is from the population in northern Virginia? She is longer than 99% of the original population, but not longer than 99.8%. What about a female who was 10.0 mm long? She's longer than 95% of known females. What about a female 9.0 mm long?

We should be aware that there is really no clear cut-off in this length problem. All we can say is that it is unlikely that a 12-mm female would be found in our population. It is also unlikely that the 11-mm female and the 10-mm female would be in our population, although it is more likely that they were than was the case for the 12-mm individual.

We can use the idea of the probability of an event occurring to decide whether to believe in a conclusion. We could conclude that the 11-mm female was not drawn from our population because the probability is 0.002 (2 chances in 1000). The probability statement specifies our confidence. The smaller the probability, the greater our confidence. For example, if the probability of an event occurring is 0.05, we are 95% (1/0.05 expressed as a percent) confident that it will not occur. How confident do we want to be? The answer to this depends on the consequences of being wrong. For our purposes, we will use a confidence level of 95% and reject any null hypothesis that has a probability level of 0.05 or less. This probability level is commonly used in many fields, but you should be aware of the fact that it is arbitrarily chosen (even if it is conventional), and may not be suitable in all cases. We might be content to be less confident (say 90%) when evaluating differences between brands of dog food, and demand more confidence (hopefully at least 99.999%) when evaluating the safety of nuclear power stations.

Back to the female who was 9 mm long. She is no different from most of the known females. We cannot reject the hypothesis that she came from our population. But we cannot know if she came from another, similar, population. Our inferences are limited in this case.

There is no certainty in statistical analysis. Unlikely things do happen, but only rarely. Our aim is to describe what we did, what we found, and what we think it means. We may differ on interpretation, but we will understand more clearly the basis for our differences, at least at our stated probability level. Since a probability level of 0.05 (or a confidence level of 95%) is widely used, we have chosen to use this level in all of the following analyses. This will greatly simplify some calculations and all of the tables in this text. Other confidence levels are available, and can be found in a comprehensive statistics text.

Selected Statistical Tests

The following section presents several statistical analyses that are appropriate for the kinds of studies presented in this text. We want to emphasize that this is only a brief introduction to selected analyses. More thorough treatments are available in statistics texts.

The following standard tables are referred to throughout the statistical material and appear on the pages indicated.

The Mann-Whitney U *Test*

APPROPRIATE USES OF THE ANALYSIS

Suppose we have two distributions of numbers and want to know whether they differ in their locations. If we measure the locations as median values, we can compare them using the Mann-Whitney U statistic. This nonparametric test compares the medians of two distributions, and is analogous to a t test or to an analysis of variance that compares two means from normal distributions.

CALCULATIONS

a. Order the combined sets of data in sequence from smallest to largest. Assign a rank to each observation. If two or more observations have the same value, then each receives the average of the ranks. For example, if two values are tied for ranks 5 and 6, then each observation is assigned the rank 5.5.

b. Sum the ranks for each sample.
R_1 = the sum of ranks for the first sample,
R_2 = the sum of ranks for the second sample, and

$$R_1 = [(N_1 + N_2)(N_1 + N_2 + 1)]/2 - R_2$$

where N_1 and N_2 are the number of observations in the first and second samples respectively. Note that the *ranks*, not the observations themselves, are used in subsequent calculations.

c. Calculate U_1 and U_2, where

$$U_1 = (N_1)(N_2) + [(N_1)(N_1 + 1)]/2) - R_1$$

and

$$U_2 = (N_1)(N_2) + [(N_2)(N_2 + 1)]/2) - R_2$$

As a check,

$$U_1 = (N_1)(N_2) - U_2$$

d. The test statistic U is the larger of the two values U_1 or U_2.
To test the null hypothesis that the two medians are equal against the alternative that they are not equal, compare U to the values in the table of critical values for the Mann-Whitney test (Table 1). If the calculated value is greater than or equal to the table value for the appropriate sample sizes, the null hypothesis is rejected at $p = 0.05$.
To test either of the two possible one-sided null hypotheses, compare U to the tabulated critical values for a one-sided test. The null hypothesis is rejected if the calculated U is larger than the tabulated value.

Table 1. Critical values for the two-sample Mann-Whitney U test.*

For Two-Sided Tests

N_2 \ N_1	3	4	5	6	7	8	9	10	11	12	13	14	15	16	17	18	19	20
1																		
2						16	18	20	22	23	25	27	29	31	32	34	36	38
3			15	17	20	22	25	27	30	32	35	37	40	42	45	46	50	52
4		16	19	22	25	28	32	35	38	41	44	47	50	53	57	60	63	66
5			23	27	30	34	38	42	46	49	53	57	61	65	68	72	76	80
6				31	36	40	44	49	53	58	62	67	71	75	80	84	89	93
7					41	46	51	56	61	66	71	76	81	86	91	96	101	106
8						51	57	63	69	74	80	86	91	97	102	108	114	119
9							64	70	76	82	89	95	101	107	114	120	126	132
10								77	84	91	97	104	111	118	125	132	138	145
11									91	99	106	114	121	129	136	143	151	158
12										107	115	123	131	139	147	155	163	171
13											124	132	141	149	158	167	175	184
14												141	151	160	169	178	188	197
15													161	170	180	190	200	210
16														181	191	202	212	222
17															202	213	224	235
18																225	236	248
19																	248	261
20																		273

For One-Sided Tests

N_2 \ N_1	3	4	5	6	7	8	9	10	11	12	13	14	15	16	17	18	19	20
1																	10	20
2			10	12	14	15	17	19	21	22	24	25	27	29	31	32	34	36
3	9	12	14	16	19	21	23	26	28	31	33	35	38	40	42	45	47	49
4		15	18	21	24	27	30	33	36	39	42	45	48	50	53	56	59	62
5			21	25	29	33	36	39	43	47	50	54	57	61	65	68	72	75
6				29	34	38	42	46	50	55	59	63	67	71	76	80	84	88
7					38	43	48	53	58	63	67	72	77	82	86	91	96	101
8						49	54	60	65	70	76	81	87	92	97	103	108	113
9							60	66	72	78	84	90	96	102	108	114	120	126
10								73	79	86	93	99	106	112	119	125	132	138
11									87	94	101	108	115	122	130	137	144	151
12										102	109	117	125	132	140	148	156	163
13											118	126	134	143	151	159	167	176
14												135	144	153	161	170	179	188
15													153	163	172	182	191	200
16														173	183	193	203	213
17															193	204	214	225
18																215	226	237
19																	238	250
20																		262

(Modified from Rohlf, F.J. and R.R. Sokal. 1969. *Statistical Tables*, W.H. Freeman.)
*The values given are for the larger U value. To be significant at $p = 0.05$, the value must be larger than the tabulated value, where N_2 is the sample with fewer observations and N_1 is the sample with the greater number of observations.

e. The accompanying Mann-Whitney table will only accommodate sample sizes up to 20. For larger samples, use U to determine a new value Z.

$$Z = \frac{U - [(N_1)(N_2)]/2}{\sqrt{[(N_1)(N_2)(N_1 + N_2 + 1)]/12}}$$

Z is actually the deviation of an observation from the mean in a normal distribution with an mean of 0.0 and a standard deviation of 1.0.

f. To test the null hypothesis that the two medians are equal against the alternative that they are not equal, compare the calculated Z value with 1.96. If Z is greater than 1.96, the null hypothesis is rejected at $p < 0.05$.

To test the null hypothesis that the median of the first sample is greater than or equal to that of the second, compare the calculated Z value with 1.64. If Z is greater than 1.64, the null hypothesis is rejected at $p < 0.05$.

To test the other one sided alternative null hypothesis that the median of the first sample is less than or equal to that of the second, compare the calculated Z with -1.64. If Z is less than -1.64, the null hypothesis is rejected at $p < 0.05$.

RATIONALE

What we have done is arrange our observations in a linear sequence. We have used ranks to indicate the overlap between distributions. If there is little overlap, then we should not reject our hypothesis.

There is intuitive sense to the form of the analysis, since as the degree of overlap decreases the distributions move apart—they become less alike.

We should note that because we have converted the observations to ranks, the scale of measurement is irrelevant. If the scale is in logarithms, or if the largest value is 100 times larger than the next largest value, those differences do not affect the ranks assigned to those observations. The ranks deal with the sequence of observations, not the magnitude of difference.

EXAMPLE

a. We conducted an experiment in which we allowed female mottled sculpins to choose between a large male and a small male. We determined female preference by recording which male she spawned with. The results of the first ten experiments are given below. Since past analyses suggested females prefer larger males, we wish to test the null hypothesis that the median of the first sample is less than or equal to the second.

b. Rank all of the observations in sequence and sum the ranks R_1 and R_2 for each sample.

WINNERS (LENGTH IN MM)	RANK	LOSERS (LENGTH IN MM)	RANK
94	11.0	82	2.5
103	17.5	90	7.5
91	9.5	104	19.0
98	13.0	75	1.0
100	15.0	88	6.0
98	13.0	84	4.0
103	17.5	90	7.5
85	5.0	98	13.0
101	16.0	82	2.5
91	9.5	105	20.0

$$R_1 = 11 + 17.5 + \ldots 9.5 = 127$$

$$R_2 = 83$$

Note that three 98's were tied for ranks 12, 13, and 14. Each was assigned a rank of 13. Also note we divided the sum of the ranks by two whenever two observations were tied for the same rank.

Recalling that

$$R_1 = [(N_1 + N_2)(N_1 + N_2 + 1)]/2 - R_2$$

we can check our sums ($N_1 = N_2 = 10$ in this case):

$$
\begin{aligned}
R_1 &= [(10 + 10)(10 + 10 + 1)]/2) - 83 \\
&= [(20)(21)]/2) - 83 \\
&= 210 - 83 \\
&= 127
\end{aligned}
$$

c. Calculate U_1 and U_2:

$$
\begin{aligned}
U_1 &= 10)(10) + (10)(11)/2] - 127 \\
&= 28
\end{aligned}
$$

and

$$
\begin{aligned}
U_2 &= (10)(10) + [(10)(11)/2] - 83 \\
&= 72
\end{aligned}
$$

d. Compare U (72) with the tabulated value for sample sizes of 10 and 10. Since 72 is less than 73, we cannot reject the hypothesis that winners and losers had equivalent median lengths, even though the median for winners was almost 10 mm greater than that for the losers in our (admittedly) small sample.

Student's t *Test*

APPROPRIATE USES OF THE ANALYSIS

Suppose that we have two samples of numbers which we know or assume to come from a normally-distributed population. In many cases, we are interested in knowing whether or not the means for the two distributions are different. We can compare the two sample means and use the *t* test to answer this question.

There are several types of *t* tests, each designed for specific applications. Here we will use a version of this test that assumes the variances for each of the two distributions being examined are equal. If this assumption cannot be met we recommend the Mann-Whitney *U* test already described.

CALCULATIONS

a. Calculate the mean ($\overline{Y}$) and variance (S^2) of each sample.

$$\overline{Y}_1 = \Sigma Y_{1i}/N_1$$

$$\overline{Y}_2 = \Sigma Y_{2i}/N_2$$

$$S_1^2 = \frac{\Sigma Y_{1i}^2 - (\Sigma Y_{1i})^2/N_1}{N_1 - 1}$$

$$S_2^2 = \frac{\Sigma Y_{2i}^2 - (\Sigma Y_{2i})^2/N_2}{N_2 - 1}$$

b. Calculate a pooled estimate of the standard deviation (Sp).

$$\mathrm{Sp} = \sqrt{\frac{(N_1 - 1)S_1^2 + (N_2 - 1)S_2^2}{N_1 + N_2 - 2}}$$

c. Calculate the *t* statistic.

$$t = \frac{(\overline{Y}_1 - \overline{Y}_2)}{\mathrm{Sp}\sqrt{1/N_1 + 1/N_2}}$$

d. Calculate degrees of freedom.

$$\mathrm{df} = N_1 + N_2 - 2$$

e. To test the null hypothesis that the two population means are equal against the alternative that they are not equal, compare the calculated *t* value with that in Table 2 for the appropriate degrees of freedom and a significance level of $p < 0.05$. Reject the null hypothesis if the absolute value of the calculated *t* is greater than the value in the table.

f. To test such a null hypothesis against the alternative that the mean of the first population is greater than that of the second, compare the calculated value of t with that given in Table 3 for the appropriate degrees of freedom. If the calculated t is greater than the table value, the null hypothesis is rejected at $p < 0.05$.

Similarly, to test the null hypothesis against the alternative that the first mean is less than that of the second, we reject the null if the calculated t is less than the negative of the tabulated t value in table 3. (n.b., In these two one-sided analyses, the sign of the calculated t value is not ignored, unlike the two-sided case given in Paragraph e).

Table 2. t values for evaluating the null hypothesis of "no difference" vs. the alternative of "difference" among two means at the significance level of 0.05.

df	*t*	df	*t*	df	*t*	df	*t*	df	*t*
5	2.57	11	2.20	17	2.11	23	2.07	29	2.04
6	2.45	12	2.18	18	2.10	24	2.06	30	2.04
7	2.36	13	2.16	19	2.09	25	2.06	40	2.02
8	2.31	14	2.14	20	2.09	26	2.06	60	2.00
9	2.26	15	2.13	21	2.08	27	2.05	120	1.98
10	2.23	16	2.12	22	2.27	28	2.05	inf.	1.96

(Modified from Rohlf, F.J. and R.R. Sokal. 1969. *Statistical Tables*, W.H. Freeman)

Table 3. t values for evaluation of one sided null hypothesis of either "greater than or equal" or "less than or equal" at the significance level of 0.05.

df	*t*	df	*t*	df	*t*	df	*t*	df	*t*
5	2.02	11	1.80	17	1.74	23	1.72	29	1.70
6	1.94	12	1.78	18	1.73	24	1.71	30	1.70
7	1.90	13	1.77	19	1.73	25	1.71	40	1.68
8	1.86	14	1.76	20	1.73	26	1.71	60	1.67
9	1.83	15	1.75	21	1.72	27	1.70	120	1.66
10	1.81	16	1.75	22	1.72	28	1.70	inf.	1.65

(Modified from Rohlf, F.J. and R.R. Sokal. 1969. *Statistical Tables*, W.H. Freeman)

RATIONALE

We can always calculate an estimate of variance. It doesn't matter if we are dealing with sets of observations or with sets of means of observations. All we need to do is calculate the appropriate variance.

$$\text{The variance of a mean} = s^2/N$$

Since our null hypothesis assumes that the samples were drawn from a common distribution, we can better estimate the variance of that distribution by pooling the data. In calculating the denominator of the t statistic, we are pooling the variances of the means.

The t test itself can be phrased as follows: What is the likelihood of drawing two means as different as these from the same population of means? This is the female soldier beetle question (introduced earlier) rephrased to deal with two observations which are means rather than single observations. The greater the difference, the less likely they come from the same population.

EXAMPLE

In an analysis of the mate preferences of male guppies (*Poecilia reticulata*), males were presented with pairs of females that had the same lengths. One of the females was virginal, having been raised in isolation from males, and the other was pregnant. The number of seconds during a 10-minute observation period when the male actively courted each female was recorded on a stopwatch. Because we might expect males to spend more time courting females who are not yet mated, our null hypothesis is that the mean time for mated females was greater than or equal to that for virgins.

a. Calculate the sums of Y_1 and Y_2, and the sums of Y_1^2 and Y_2^2.

TRIAL	VIRGIN	MATED
1	46	53
2	27	7
3	58	2
4	71	12
5	109	88
6	80	32
7	61	0
8	6	9
9	32	23
10	48	55
11	25	42
12	42	10
13	38	7
14	20	0

$\Sigma Y_1 = 663$ $\qquad \Sigma Y_2 = 349$

$\Sigma Y_1^2 = 40849$ $\qquad \Sigma Y_2^2 = 16322$

$N_1 = 14$ $\qquad N_2 = 14$

b. Calculate the means for each sample.

$$\bar{Y}_1 = 663/14 = 47.4$$

$$\bar{Y}_2 = 340/14 = 24.3$$

c. Calculate the variance for each sample.

$$s_1^2 = [40849 - (663)^2/14]/13 = 727.0$$

$$s_2^2 = [17322 - (340)^2/14]/13 = 697.3$$

d. Calculate the standard deviation.

$$Sp = \sqrt{\frac{(13)727.0 + (13)697.3}{26}} = 26.7$$

e. Calculate the t value.

$$t = \frac{47.4 - 24.3}{26.7\ (0.4)} = 2.32$$

f. Calculate degrees of freedom.

$$df = 14 + 14 - 2 = 26$$

g. Compare the calculated t value with that having 26 degrees of freedom in Table 3, remembering that we have decided to examine a one-sided hypothesis. Since the calculated t (2.32) is greater than the table value of 1.71, we reject the null hypothesis and conclude that we do have evidence that males directed more attention to virgin females.

Analysis of Variance

APPROPRIATE USES OF THE ANALYSIS

In spite of its name, the analysis of variance is a test for differences among population means. Like the t test, it compares means. Unlike the t test, it can compare any number of means. Suppose, for example, we counted the number of different types of pollen grains collected by bees during a single foraging trip. If we examined only two species of bees, we might compare the mean number of pollen types using a t test. If we examined three or four species of bees, we might compare mean numbers using an analysis of variance. In statistical terms, each bee species would be a "treatment" and the set of observations for each bee species would be a "group."

This is a parametric procedure, and, like the t test, assumes that populations are normally distributed and have equal variances. Because biological data often meet these assumptions, this analysis is frequently appropriate.

In the analysis of variance, the null hypothesis is always that all populations being compared have the same mean. The alternative hypothesis is that at least one of the means is different. Discovering which of the means is different is beyond the scope of this text, but methods are available, and can be found in many statistics texts.

CALCULATIONS

a. Calculate the sums of Y_i and the sums of squared Y_i's for each sample.

b. Calculate the grand sum, the grand sum of squared Y_i's, and the total sample size.

Grand sum: $\Sigma\Sigma Y_i$ (the sum of all the observations combined)

Grand sum of squared observations: $\Sigma\Sigma Y_i^2$ (the sum of all the squared observations)

Total sample size: ΣN_i

Number of treatments (groups compared): c
Sample sizes for each of the c groups: N_i

c. Calculate the correction term (CT).

$$\text{CT} = [(\Sigma\Sigma Y_i)^2/\Sigma N_i]$$

d. Calculate the total sum of squares.

$$\text{Total} = \Sigma\Sigma Y_i^2 - \text{CT}$$

e. Calculate the treatment sum of squares.

$$\text{Treatment} = \Sigma[(\Sigma Y_i)^2/\Sigma N_i] - \text{CT}$$

f. Calculate the error sum of squares.

$$\text{Error} = \text{Total} - \text{Treatment}$$

g. Complete the ANOVA table.

SOURCE OF VARIATION	df	SUM OF SQUARES	MEAN SQUARE (MS)	F
Among treatments	$c - 1$	Treatment	Treatment/$(c - 1)$	MS Treatment/MS error
Within groups	$\Sigma N_i - c$	Error	Error/$\Sigma N_i - c$	
Total	$\Sigma N_i - 1$	Total		

h. Compare the calculated F value with the value in Table 4 for $(c - 1)$ numerator and $\Sigma (N_i - c)$ denominator degrees of freedom. If the calculated F is larger than the one in the table, the null hypothesis of equal means is rejected.

i. The ratio of Treatment Sum of Squares/Total Sum of Squares (n.b., not mean squares) indicates the proportion of the total variability among observations that is explained by the treatments. This value is called the **coefficient of determination.** (For another application of this coefficient, see the following discussion of regression analysis.) Note that it is only appropriate to calculate the coefficient of determination if you found a significant difference among treatments.

RATIONALE

When performing an analysis of variance we calculate two estimates of the population variance. The error mean square (MS) in the ANOVA table is an estimate of the population variance whether the null hypothesis is true or not (that is, whether all means are equal or not). If the null hypothesis is true, then the treatment mean square in the table also estimates the common population variance. Thus treatment MS divided by error MS will be a small number and average unity. If the null hypothesis is not true, that is if some means are different, the treatment MS will be larger than the population variance. The ratio of treatment MS to error MS will therefore be much larger than one.

The F statistic is simply the ratio of two variances. In this case it measures similarity between the variability among treatments and the variability within groups.

Table 4. Critical values for the F statistic at the 0.05 probability level.

DENOMINATOR (df)	NUMERATOR DEGREES OF FREEDOM (df)															
	1	2	3	4	5	6	7	8	9	10	12	15	30	60	120	Inf.
1	161.0	200.0	216.0	225.0	230.0	234.0	237.0	239.0	241.0	242.0	244.0	246.0	250.0	252.0	253.0	254.0
2	18.5	19.0	19.2	19.3	19.4	19.4	19.4	19.4	19.4	19.4	19.4	19.4	19.5	19.5	19.5	19.5
3	10.1	9.8	9.3	9.1	9.0	8.9	8.9	8.8	8.8	8.8	8.7	8.7	8.6	8.6	8.6	8.5
4	7.7	6.9	6.6	6.4	6.3	6.2	6.1	6.0	6.0	6.0	5.9	5.9	5.8	5.7	5.7	5.6
5	6.6	5.8	5.4	5.2	5.0	5.0	4.9	4.8	4.8	4.7	4.7	4.6	4.5	4.4	4.4	4.4
6	6.0	5.1	4.8	4.5	4.4	4.3	4.2	4.1	4.1	4.1	4.0	3.9	3.8	3.7	3.7	3.7
7	5.6	4.7	4.4	4.1	4.0	3.9	3.8	3.7	3.7	3.6	3.6	3.5	3.4	3.3	3.3	3.2
8	5.3	4.5	4.1	3.8	3.7	3.6	3.5	3.4	3.4	3.4	3.3	3.2	3.1	3.0	3.0	2.9
9	5.1	4.3	3.9	3.6	3.5	3.4	3.3	3.2	3.2	3.1	3.1	3.0	2.9	2.8	2.8	2.7
10	5.0	4.1	3.7	3.5	3.3	3.2	3.1	3.1	3.0	3.0	2.9	2.8	2.7	2.6	2.6	2.5
12	4.8	3.9	3.5	3.3	3.1	3.0	2.9	2.8	2.8	2.8	2.7	2.6	2.5	2.4	2.3	2.3
15	4.5	3.7	3.3	3.1	2.9	2.8	2.7	2.6	2.6	2.5	2.5	2.4	2.2	2.2	2.1	2.1
30	4.2	3.3	2.9	2.7	2.5	2.4	2.3	2.3	2.2	2.2	2.1	2.0	1.8	1.7	1.7	1.6
60	4.0	3.2	2.8	2.5	2.4	2.2	2.2	2.1	2.0	2.0	1.9	1.8	1.6	1.5	1.5	1.4
120	3.9	3.1	2.7	2.5	2.3	2.2	2.1	2.0	2.0	1.9	1.8	1.8	1.6	1.4	1.4	1.2
Inf	3.8	3.0	2.6	2.4	2.2	2.1	2.0	1.9	1.9	1.8	1.8	1.7	1.5	1.3	1.2	1.0

(Modified from Rohlf, F.J. and R.R. Sokal. 1969. *Statistical Tables*. W.H. Freeman.)

EXAMPLE

Forked fungus beetles (*Bolitotherus cornutus*) are small beetles that eat various species of shelf fungus. The males of this species have horns with which they fight, rather like tiny bison, and males' horns come in different sizes. In order to examine the effects of larval food supply on the size of an adult male's horns, we raised larvae on three common fungal food sources under identical conditions. When the adults emerged we measured their horn lengths, which are reported here in occular micrometer units.

a. To save space, we report the appropriate sums rather than all of the data:

GROUP INFORMATION	TREATMENT (FOOD SOURCE)		
	Ganoderma tsugae	*G. applanatum*	*Fomes fomentarius*
ΣY_i	413	174	53
ΣY_i^2	10155	2980	261
N	25	17	12

b. Calculate the grand sum, the grand sum of the squared observations, and the grand sample size.

$$\Sigma\Sigma Y_i = 640$$
$$\Sigma\Sigma Y_i^2 = 13396$$
$$\Sigma N_i = 54$$

c. Calculate the correction term.

$$CT = (640)(640)/54 = 7585.19$$

d. Calculate the total sum of squares.

$$\text{Total} = 13396 - 7585.19 = 5810.81$$

e. Calculate the treatment sum of squares.

$$\text{Treatment} = [(413)(413)/25 + (174)(174)/17 + (53)(53)/12] - 7585.19$$
$$= 8837.78 - 7585.19$$
$$= 1252.59$$

f. Calculate the error sum of squares.

$$\text{Error} = 5810.81 - 1252.59 = 4558.22$$

g. Complete the ANOVA table.

SOURCE OF VARIATION	df	SUM OF SQUARES	MEAN SQUARE	*F*
Among treatments	2	1252.59	626.30	7.01
Within groups	51	4558.22	89.38	
Total	53	5810.81		

h. Compare the calculated F statistic with the critical value given in Table 4 with 2 and 51 degrees of freedom. Although we might interpolate the critical F values for the exact degrees of freedom, our calculated value of 7.01 is much larger than the closest appropriate value in the table (for 2 and 60 df, F = 4.2); thus we have sufficient reason to reject the null hypothesis that the three means were equal. This means that the type of food the larvae ate had an effect on the horns of the adult male.

i. Since we did find an effect of food type (i.e., an effect ascribable to the treatment), we can proceed to calculate the coefficient of determination:

$$\text{Treatment SS/Total SS} = 1252.59/5810.81 = 0.22$$

Thus we have attributed 22% of the variation among male horn lengths to larval diet effects.

Comparison of Variances

APPROPRIATE USES OF THE ANALYSIS

Suppose we have two sets of observations and wish to know whether one is more (or less) variable than the other. We might want to know whether population variances are different before proceeding with some other inferential test (like a *t* test). Alternatively, we might be interested in the biological significance of different variances. For example, if stabilizing selection were operating in a population of beetles, we might expect that mating individuals would be less variable in body size than nonmating individuals. Regardless of the reasons for our interest, the *F* test is a very simple parametric analysis that compares variances. This analysis assumes that the samples were taken from normal distributions. For this test the null hypothesis is always that variances are equal.

CALCULATIONS

a. Calculate the variance for each sample.

b. Calculate *F* by dividing the larger variance by the smaller variance.

c. Compare the result with the tabulated *F* value with (numerator $N - 1$) and (denominator $N - 1$) degrees of freedom (Table 4).

d. If the calculated value is greater than the tabulated value reject the null hypothesis that the variances are equal.

RATIONALE

An *F* distribution is a probability distribution for ratios between variances (it is not unlike a normal distribution in that regard). So it calculates the probability of getting ratios of a certain magnitude, given that the samples were drawn at random from a common population. Clearly, as the ratio increases, the less likely we are to have found that ratio under the assumption that the variances are estimates of a common variance.

EXAMPLE

a. In the example given for the *t* test, we compared the responses of male guppies (*Poecilia reticulata*) to virgin and pregnant females. We wanted to know whether males spent more time courting the virgins, and compared mean courtship times for each type of female. An assumption of that analysis was that courtship times for virgin and mated females had the same variance. In this example we will test the validity of that assumption. Since the actual data have already been presented, we simply give sample sizes and variances here.

	VIRGIN FEMALES	MATED FEMALES
N	14	14
S^2	727.0	697.3

b. Calculate the F statistic as the ratio of the larger to the smaller variance.

$$F = 727.0/697.3 = 1.1$$

c. Compare the calculated F (1.1) with the tabulated value of F having (14 − 1 = 13 numerator degrees of freedom) and (14 − 1 = 13 denominator degrees of freedom) in Table 4. Since the calculated value is much smaller than the closest table value ($F_{15/15} = 2.4$) we conclude that there is not sufficient evidence to reject the null hypothesis of equality. Thus the variance in courtship times for virgin females was equal to that for mated females.

Note that the F test is designed to compare two variances and no more. We can use it when testing the assumption of equal variances prior to comparing means with a t test. We cannot use the F test when testing the assumption that more than two variances are equal prior to comparison of more than two means with an analysis of variances (ANOVA). Special tests for differences among several variances are available, and can be found in any comprehensive statistics text.

Correlation Analysis: Spearman Rank Correlation

APPROPRIATE USES OF THE ANALYSIS

Suppose we have two distributions of numbers, and we wish to examine the interdependence of these two distributions (i.e., we want to know how the two types of variables covary, or vary together). Correlation analysis allows such examination. It allows us to evaluate the degree to which two variables vary or change together. It does not allow us to evaluate causation; that is, it will tell us if two variables change in the same direction at the same time, it will not tell us if one variable causes the other to change. One may cause the other to change, or both may change because of some third, unmeasured, variable.

Three general patterns of joint variation can be imagined. Both variables may increase at the same time, one may increase while the other decreases, or there may be no joint variation at all; that is, the two variables are completely independent of each other. When a positive relationship exists between the two variables, we will find a positive correlation. When one variable increases while the other decreases, we will find a negative correlation. When the two are independent, we will find no correlation.

Both the sign (+ or —) and the tendency of two variables to change together can be summarized in a single quantity called a **correlation coefficient.** This will be a number that varies from —1 to +1. A correlation coefficient of +1 indicates that the two variables always change together and in the same way. As one grows larger, the other does too, and there is no unexplained variability (or error) in the relationship. A correlation of —1 indicates a perfect negative relationship, in which the variables change together, but in opposite directions. A correlation of 0 indicates no joint dependence. Any value between 0 and +1 indicates a positive correlation that is less than perfect: as one variable gets bigger, so does the other—but with variability, not with a one-to-one correspondence. Similarly, any correlation coefficient between 0 and —1 shows an imperfect joint variation in a negative relationship.

There are several types of correlation coefficients. All are designed to do the same thing: describe the degree of covariation among two sets of variables in a sample. Since we want to infer the true covariation among these variables in the statistical population, we will always test the null hypothesis that the true correlation is zero. Here we will use a nonparametric coefficient called the Spearman's rank correlation coefficient. It is usually abbreviated r_S.

CALCULATIONS

a. Rank each observation in the first distribution from smallest to largest. For example, assign the value of 1 to the smallest observation in the first distribution, the value 2 to the second smallest, and so forth up to the *n*th or last observation. Assign average ranks when ties occur.

b. Repeat the ranking procedure for the second distribution. Use the same procedures described above.

c. Find the difference in ranks for each pair of observations.

d. Square each of the differences in ranks.

e. Calculate

$$r_S = 1 - \frac{6(\Sigma \text{differences}^2)}{(N)(N-1)(N+1)}$$

where N is the number of paired variables.

f. To test the null hypothesis that the true correlation is not different from zero, compare the calculated r with the tabulated value for the appropriate sample size (Table 5).

Table 5. Critical values for Spearman's rank correlation. If r_S is greater than the tabulated value, the null hypothesis is rejected at $p < 0.05$.

N	r_S	N	r_S	N	r_S
5	1.000	16	.503	27	.382
6	.866	17	.485	28	.375
7	.786	18	.472	29	.368
8	.738	19	.460	30	.362
9	.700	20	.447	35	.335
10	.648	21	.433	40	.313
11	.618	22	.425	50	.279
12	.587	23	.415	60	.255
13	.560	24	.406	70	.235
14	.538	25	.398	80	.220
15	.521	26	.390	100	.197

(Modified from Zar, J.H. 1974. *Biostatistical Analysis*, Prentice-Hall.)

RATIONALE

The rationale of Spearman's coefficient can be understood by examining the formula used to calculate it. If both sets of variables increase, then the two sets of ranks will be equivalent, the sum of the squared differences in ranks will be small, and r will be large. If the rankings are identical, the coefficient will be 1.0. If one set of variables increases while the other decreases, the sum of the squared differences in rank will be large. If the ranks are exactly opposite, the coefficient will be −1.0.

EXAMPLE

The following data represent the weights in grams of male and female ambush bugs (Phymatidae) captured in copula.

PAIR	MALE	RANK	FEMALE	RANK	$\lvert R - R \rvert$	$(R - R)^2$
1	.0064	8.0	.0135	12.0	4.0	16.00
2	.0071	10.5	.0118	9.5	1.0	1.00
3	.0049	1.0	.0105	4.5	3.5	12.25
4	.0056	4.0	.0110	6.5	2.5	6.25
5	.0051	2.0	.0105	4.5	2.5	6.25
6	.0083	13.0	.0107	8.0	5.0	25.00
7	.0071	10.5	.0104	3.0	7.5	56.25
8	.0065	9.0	.0097	2.0	7.0	49.00
9	.0062	7.0	.0120	11.0	4.0	16.00
10	.0075	12.0	.0143	13.0	1.0	1.00
11	.0053	3.0	.0084	1.0	2.0	4.00
12	.0060	5.5	.0108	9.5	4.0	16.00
13	.0060	5.5	.0110	6.5	1.0	1.00

$$r_S = 1 - \frac{(6)(210)}{(12)(13)(14)}$$

$$= 0.423$$

Since the calculated coefficient (0.42) is not larger than the tabulated value (0.56) or a sample size of 13 (Table 5), we conclude that there is not a significant rank correlation between male and female weights at $p < 0.05$. Note that the same calculated r_S (0.42) would have been significant for a larger sample size (>22). This illustrates the fact that true but small correlations (i.e., some covariance, but not much) cannot be detected with small samples. In this case, we cannot support the argument that male and female sizes covary, but we might be tempted to collect more data.

Correlation Analysis: Pearson's r

APPROPRIATE USES OF THE ANALYSIS

Pearson's r is the parametric equivalent of the Spearman r_S coefficient. It too summarizes joint variation in a single number that is greater than or equal to −1 and less than or equal to +1. As a parametric statistic, it does assume that the distributions of the variables involved are normal. This analysis further assumes that the relationship between the two variables is linear. Despite the additional assumptions, Pearson's r may be much easier to calculate for large sample sizes where the ranking necessary for the Spearman coefficient would be extremely tedious. Furthermore, many pocket calculators are equipped for the calculations involved.

CALCULATIONS

a. Calculate the sums, and sums of squared observations, for each variable.

b. Calculate the sum of the products of Y_1 and Y_2, $\Sigma Y_1 Y_2$.

c. Calculate the Σy_1^2, Σy_2^2, and $\Sigma y_1 y_2$.

$$\Sigma y_1^2 = \Sigma Y_1^2 - (\Sigma Y_1)^2/N$$
$$\Sigma y_2^2 = \Sigma Y_2^2 - (\Sigma Y_2)^2/N$$
$$\Sigma y_1 y_2 = \Sigma Y_1 Y_2 - [(\Sigma Y_1)(\Sigma Y_2)/N]$$

d. Calculate r.

$$r = \Sigma y_1 y_2 / \sqrt{(\Sigma y_1^2)(\Sigma y_2^2)}$$

e. To test the null hypothesis that the true correlation is not different from zero, compare the calculated value with the tabulated r value with the appropriate sample size (= the number of pairs of observations; Table 6). If $|r|$ is greater than the tabulated value, the null hypothesis is rejected at $p < 0.05$.

RATIONALE

Pearson's r requires that we calculate a new quantity, $\Sigma y_1 y_2$. If we examine how this is calculated we will note that it is simply the numerator of a variance estimate. If we divide it by $N - 1$ we get a true variance, called the **covariance**.

Pearson's r is the ratio of the covariance to an estimate of the independent variances of the two samples. If the two variables increase together, then r will approach +1.0. That is, the covariance will equal the square root of the product of the two independent variances. Like Spearman's rank correlation, Pearson's rank correlation can never be greater than +1.00, nor smaller than −1.00.

Table 6. Critical values for Pearson's *r*.

SAMPLE SIZE	*r*	SAMPLE SIZE	*r*
3	.997	20	.444
4	.950	25	.396
5	.878	30	.361
6	.811	35	.334
7	.754	40	.310
8	.707	45	.295
9	.666	50	.279
10	.632	100	.194
12	.576	250	.125
14	.532	500	.089
16	.497	1000	.062
18	.468		

(Modified from Rohlf, F.J. and R.R. Sokal. 1969. *Statistical Tables*, W.H. Freeman.)

EXAMPLE

In October of 1984 we studied the placement of larval cases by evergreen bagworms (*Thyridopteryx ephemeraeformis*) on black locust trees (*Robinia pseudoacacia*). We measured the length of each case and the diameter of the twig to which it was attached. We want to know whether there is any correlation between these two variables. (Data are given in mm.)

TWIG DIAMETER	CASE LENGTH	TWIG DIAMETER X CASE LENGTH
0.9	49.0	44.10
0.9	39.5	35.55
1.1	31.2	34.32
0.8	39.5	31.60
3.6	50.0	180.00
5.5	53.0	291.50
4.9	59.2	290.08
0.8	33.2	26.56
1.28	46.1	59.01
1.1	43.9	48.28
2.38	33.8	80.44
7.8	49.7	387.66

a. Calculate the sums of Y_1, Y_2, Y_1^2, Y_2^2, and the sum of the products of Y_1Y_2

$\Sigma Y_1 = 31.06$ $\Sigma Y_2 = 528.10$ $\Sigma Y_1Y_2 = 1509.10$

$\Sigma Y_1^2 = 140.68$ $\Sigma Y_2^2 = 24075.77$ $N = 12$

b. Calculate $\sum y_1^2$, $\sum y_2^2$, and $\sum y_1 y_2$.

$$\sum y_1^2 = 140.68 - [(31.06)(31.06)/12] = 60.29$$

$$\sum y_2^2 = 24075.77 - [(528.10)(528.10)/12] = 834.97$$

$$\sum y_1 y_2 = 1509.10 - [(31.06)(528.10)/12] = 142.20$$

c. Calculate r.

$$r = 142.20/\ \sqrt{(60.29)(834.97)}$$

$$= 0.63$$

d. Compare the calculated r (0.63) with the critical value for a sample of 12 pairs of observations given in Table 6. Since our calculated value of 63 is bigger than the table value of 0.58 we reject the null hypothesis that there is no joint variation and conclude that we have evidence that twig diameter and case length were correlated.

Correlation coefficients tell us something about how the two variables covary. Pearson's differs from Spearman's in that it is quite sensitive to extreme values (this was also true in our comparison of a t test or ANOVA with the Mann-Whitney U test). As a consequence we may sometimes be drawing conclusions based on a rare event: the outlying observation. It is desirable to look at your data (i.e., plot the points and examine visually) and determine if this is a problem before you do any calculations.

Pearson's r is sensitive to curvilinearity in the data. If the two variables do not covary in a linear fashion, then Pearson's r will underestimate the correlation between the two variables. Spearman's rank correlation will be unaffected. Again, it is useful to plot your data before you begin your analysis.

When we analyze correlation we want to know two things: Is the correlation we find real (i.e., is the true correlation different from zero)? and, How tight is the relationship (i.e., how big is the coefficient)? We answer the first question by comparing our calculated r with the table of critical values (Table 6). The size of the calculated correlation coefficient provides an answer to the second question, and the square of this coefficient provides another, commonly used, answer. This squared coefficient (r^2) is the same coefficient of determination described in the discussions of analysis of variance and again in regression. Here it summarizes the proportion of variation in the variables Y_1 and Y_2 due to covariation between them.

In the example, we calculated an r of 0.63; thus the coefficient of determination (r^2) would be 0.40, or 40%. That is, only 40% of the total variation in the two variables covaries between them, and 60% is due to other causes. While this may be a significant correlation, it accounts for less than half of the variation in the samples. We should keep this distinction in mind whenever we deal with correlations. The significance of a correlation coefficient, like all other statistics, is dependent on sample size. The larger our sample, the smaller the difference we can detect. Yet those differences may not tell us much about the total picture. A correlation coefficient of 0.30 accounts for less than 10% of the total variation; we would not place any serious bets on that relationship, regardless of its significance. However, it tells us something about tendencies, even though they are rather weak.

Partial Correlation Coefficients

APPROPRIATE USES OF THE ANALYSIS

Suppose we have measured several attributes of females that may influence the number of eggs that a female produces. These might include size and age. When we calculate the correlation between each of these variables and egg number, we may find significant correlations. While it is tempting to conclude that both age and size influence egg number, the conclusion may be premature since it is likely that size and age are correlated with each other. Hence, part of the variation in egg number that we attribute to size is actually a reflection of the correlation between size and age; likewise the correlation between age and egg number is partly a reflection of the correlation between size and age.

Partial correlations allow us to examine the unique correlations among variables by "holding the other variables constant." In a sense it is a form of statistical control. By using this technique we can examine (using the example above) the correlation between age and egg number while holding size constant. In a like manner, we can examine the effect of size on egg number while we hold age constant. We can then draw conclusions regarding the separate relationship each of these variables to egg number.

CALCULATIONS

a. Calculate correlations (Pearson's r) among the variables of interest.

b. For *three* variables (a, b, and c) the partial correlation coefficient is calculated as:

$$r_{ab.c} = \frac{r_{ab} - (r_{ac})(r_{bc})}{\sqrt{[1 - (r_{ac})^2][1 - (r_{bc})^2]}} \qquad [1]$$

where the dot between the subscripts is read as "the correlation between a and b, holding c constant," and r_{ab} is read simply as "the Pearson correlation between a and b."

c. For *four* variables (a, b, c, and d) partial correlation coefficients are calculated as:

$$r_{ab.cd} = \frac{r_{ab.d} - (r_{ac.d})(r_{bc.d})}{\sqrt{[1 - (r_{ac.d})^2][1 - (r_{bc.d})^2]}} \qquad [2]$$

d. For *five* variables (a, b, c, d and e) the partial correlations are calculated as follows:

$$r_{ab.cde} = \frac{r_{ab.de} - (r_{ac.de})(r_{bc.de})}{\sqrt{[1 - (r_{ab.de})^2][1 - (r_{bc.de})^2]}} \quad [3]$$

e. The equation for partial correlations can be expanded in the above pattern to as many variables as you wish. The significance of any partial correlation coefficient is evaluated by comparing it to the tabulated r value with $N - K$ degrees of freedom (Table 6), where N is the sample size and K is the number of variables involved.

f. The coefficient of determination is equal to the sum of all squared partial correlation coefficients.

RATIONALE

Consider the following diagram of correlations among three variables.

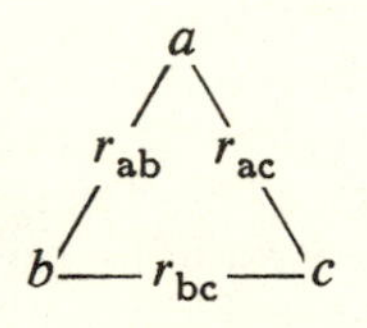

The numerator of Equation 1 subtracts the joint effects of c on a and b from the correlation of a with b. As r_{ac} and r_{bc} increase, $r_{ab.c}$ will decrease; i.e., as c has more and more effect on both a and b, the relationship between a and b independent of c becomes less and less. The denominator in Equation 1 is a correction term which takes into account the size of the confounding correlations.

EXAMPLE

The following diagram indicates the Pearson correlations between size (total length), clutch weight (weight of the entire mass of eggs), and age (years) for 64 female mottled sculpins (*Cottus bairdi*) living in the Madison River in southwestern Montana.

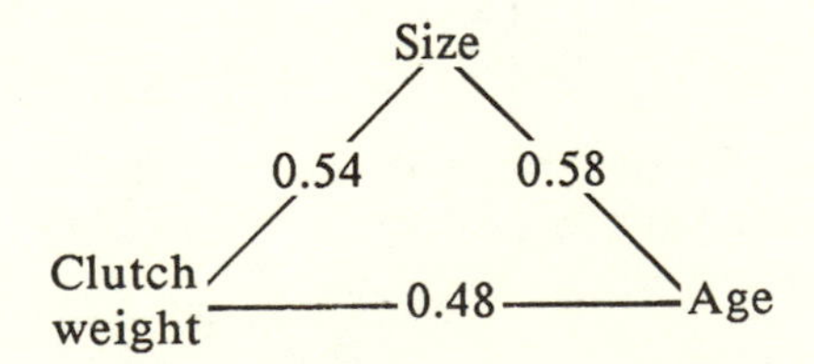

Obviously, size, clutch weight, and age are all interrelated. Analysis of partial correlations will allow us to tease out the separate relationships among all pairs of variables while holding the third variable constant.

a. Calculate the partial correlation between size and clutch weight.

$$\begin{aligned} r_{sc.a} &= 0.54 - (0.58)(0.48)/\sqrt{[1 - (0.58)^2][1 - (0.48)^2]} \\ &= (0.54 - 0.28)/\sqrt{(0.66)(0.77)} \\ &= 0.37 \end{aligned}$$

b. Compare the observed value with the tabulated r from the table of critical values for Pearson's coefficient (Table 6), using $64 - 3 = 61$ degrees of freedom. The correlation between size and clutch weight is positive and significant when female age is held constant.

c. Calculate the partial correlation between age and clutch weight.

$$r_{ac.s} = 0.48 - (0.58)(0.54)/\sqrt{[1 - (0.58)^2][1 - (0.54)^2]}$$
$$= 0.25$$

Comparison with the table of critical values indicates that this correlation is also significant.

d. Calculate the coefficient of determination.

$$= (0.37)^2 + (0.25)^2 = 0.20 \text{ or } 20\%$$

The conclusions that we draw are that 20% of the variance in total clutch weight is associated with variability in size or age of the female. Size accounts for over 70% of this variation—$(0.37)^2/[(0.37)^2 + (0.25)^2]$. Thus both size and age are significantly partially correlated with clutch weight, but size is the more important correlate for this population.

Regression Analysis

APPROPRIATE USES OF THE ANALYSIS

Suppose we measure two variables and want to know the extent to which one of them (called the dependent variable, Y_2) depends on the other (called the independent variable, Y_1). For example, we might measure the duration of courtship behaviors by male beetles having different weights, and want to know how duration depends on weight. Regression analysis allows us to describe the association between these two variables.

This analysis assumes that the mean of the dependent variable (Y_2) is a linear function of the independent variable (Y_1), and that the observed values of the dependent variable are distributed normally around their mean. It is reasonable to interpret this as a cause and effect relationship between Y_1 and Y_2. For example, suppose we analyzed our measures of courtship duration and weight and found that duration (D) depended on weight (W) in this linear fashion:

$$D = 1.0 + 3.2W$$

This means that we would predict that duration would be 7.4 when W equalled 2 (i.e., that the expected value of D, or the mean of D, would equal 7.4 when W equalled 2), and that differences in duration were caused by differences in weight.

When we calculate a regression equation, we are solving for the elements in the general equation for a straight line:

$$Y_2 = a + b(Y_1)$$

The y-intercept (a) is that value of Y_2 expected when $Y_1 = 0$. The slope (b) is the rate at which Y_2 changes with respect to Y_1. Note again that this analysis assumes that the relationship between variables is linear. If it is not, then it will be necessary to transform the data to make them linear. It is a good idea to plot the data before proceeding with this analysis.

CALCULATIONS

These calculations use the same terms described when discussing Pearson's correlation.

a. Calculate the sums, sums of square observations, and sum of cross-products.

b. Calculate the means for each sample.

c. Calculate Σy_1^2, Σy_2^2, and $\Sigma y_1 y_2$.

d. Calculate b.

$$b = \Sigma y_1 y_2 / \Sigma y_1^2$$

e. Calculate a.

$$a = \overline{Y}_1 - (b)(\overline{Y}_2)$$

f. To test the null hypothesis that the regression coefficient (b) is zero against the alternative that it is different from zero, calculate the sum of squares due to regression.

$$\text{Regression} = (\Sigma y_1 y_2)^2 / \Sigma y_1^2$$

g. Complete the ANOVA table.

SOURCE OF VARIATION	df	SUM OF SQUARES	MEAN SQUARE	F
Regression	1	$(\Sigma y_1 y_2)^2/\Sigma y_1^2$	SS/df	Regression MS/Error MS
Error	$N - 2$	by subtraction	SS/df	
Total	$N - 1$	Σy_2^2		

h. Compare the calculated F value with the tabulated value with 1 and $N - 2$ degrees of freedom (Table 4)

i. An estimate of the proportion of total variance in the dependent variable explained by its regression on the independent variable can be obtained by dividing regression sum of squares by total sum of squares. This value is the coefficient of determination, and is mathematically equivalent to r^2 (the correlation between Y_1 and Y_2 squared).

RATIONALE

The rationale for linear regression is very similar to that for the analysis of variance. The **error mean square** is an estimate of the variation of Y_2 around its mean. The **regression mean square** is an estimate of the variation of Y_2 around its mean plus the additional variation in Y_2 caused by the effect of Y_1 on the mean of Y_2.

If there is no regression (no effect of Y_1 on the mean of Y_2) then the error mean square and the regression mean square are equal, and their ratio (F) equals unity. The greater the effect of Y_1 on the mean of Y_2, the greater the ratio of regression mean square to error mean square, and the greater the calculated F statistic.

EXAMPLE

The data are measures of total leaf area and flower number for a population of *Gentiana plebia*, an annual plant found in the Rocky Mountains.

a. Calculate the sums, sums of squared observations, the means, and the sum of the cross-products.

LEAF AREA	FLOWER NUMBER	LEAF AREA	FLOWER NUMBER	LEAF AREA	FLOWER NUMBER
5.2 cm	7	4.6	8	12.3	10
5.3	7	5.7	9	5.5	12
9.6	14	3.0	9	2.6	7
3.4	6	4.8	7	2.5	7
3.5	8	1.8	2	2.2	7
.8	2	1.1	3	2.0	5
.7	6	2.6	6	2.2	8
.0	7	4.7	9	3.6	12

Leaf Area = Y_1; $\Sigma Y_1 = 91.7$, $\Sigma Y_1^2 = 515.8$, $\overline{Y_1} = 3.8$
Flower Number = Y_2; $\Sigma Y_2 = 178.0$, $\Sigma Y_2^2 = 1512.0$, $\overline{Y_2} = 7.4$
$\Sigma Y_1 Y_2 = 799.5$, Sample size = $N = 24$

b. Calculate Σy_1^2, Σy_2^2, and $\Sigma y_1 y_2$.

$$\Sigma y_1^2 = 515.8 - (91.7)(91.7)/24$$
$$= 165.4$$
$$\Sigma y_2^2 = 1512 - (178.0)(178.0)/24$$
$$= 191.8$$
$$\Sigma y_1 y_2 = 799.5 - (178.0)(91.7)/24$$
$$= 119.4$$

c. Calculate b.

$$b = 119.4/165.4$$
$$= 0.7$$

d. Calculate a.

$$a = 3.8 - (0.7)(7.4)$$
$$= -0.14$$

e. Calculate reduction sum of squares.

$$\text{Reduction} = (119.4)(119.4)/165.4$$
$$= 86.2$$

f. Complete the ANOVA table.

ANOVA for flower number and leaf area.

SOURCE OF VARIATION	df	SUM OF SQUARES	MEAN SQUARE	F
Regression	1	86.2	86.2	18.7
Residual	23	105.6	4.6	
Total	23	191.8		

g. Compare the calculated F value (18.7) with the tabulated value for 1 numerator and 23 denominator degrees of freedom (Table 4). The closest degrees of freedom in this table are 1 and 30, which yield an F of 4.5. Since the calculated value is much larger than the table value, we reject the null hypothesis that there is no effect of regression and conclude that flower number is a linear function of leaf area.

h. Calculate the coefficient of determination.

$$\begin{aligned}\text{Coefficient of determination} &= \text{Regression SS/Total SS} \\ &= 86.2/191.8 \\ &= 0.45\end{aligned}$$

We therefore conclude that about 45% of the variability in flower number is due to variation in leaf area.

The Chi-Square Test: A Priori Hypotheses

APPROPRIATE USES OF THE ANALYSIS

Suppose we have some expected distribution of events and wish to know whether or not a distribution of observed events differs significantly from that expected. The Chi-square (χ^2) test allows us to compare the observed and expected distributions and determine whether they are the same or different. An alternative analysis, which uses a statistic called G is also available. Since G is less used, we will only present χ^2 at this point.

CALCULATIONS

a. Determine the expected frequency for each event.

Expected = $(N)(P)$ where N = total sample size

P = probability of obtaining that event

b. Determine the observed frequency for each event by examining the original data.

c. Compare the observed and expected frequencies using the χ^2 statistic.

$$\chi^2 = \Sigma\,(\text{observed} - \text{expected})^2/\text{expected}$$

d. Determine the degrees of freedom associated with the test.

degrees of freedom $= a - 1$, where a is the number of categories compared.

e. To evaluate the null hypothesis of no difference between the observed and expected distributions, compare the calculated χ^2 with that in Table 7 for the appropriate degrees of freedom. If the calculated value exceeds the tabulated one, reject the null hypothesis at $p = 0.05$.

Table 7. Critical values for the χ^2 statistic at $p = 0.05$.

df	χ^2	df	χ^2	df	χ^2
1	3.8	16	26.3	35	49.8
2	6.0	17	27.6	40	55.8
3	7.8	18	28.9	45	61.7
4	9.5	19	30.1	50	67.5
5	11.1	20	31.4	55	73.3
6	12.6	21	32.7	60	79.1
7	14.1	22	33.9	65	84.8
8	15.5	23	35.2	70	90.5
9	16.9	24	36.4	75	96.2
10	18.3	25	37.6	80	101.9
11	19.7	26	38.9	85	107.5
12	21.0	27	40.1	90	113.2
13	22.4	28	41.3	95	118.8
14	23.7	29	42.6	100	124.3
15	25.0	30	43.8		

(From Rohlf, F.J. and R.R. Sokal. 1969. *Statistical Tables*, W.H. Freeman.)

RATIONALE

The rationale for this and the following chi-square analyses can best be understood by inspection of the formula used to calculate the statistic. This formula sums the squared differences between what was observed and what was expected, divided by the size of the expected. This means that the test statistic will be very small when observed and expected values are similar, and will be very large when there is little or no congruency. The actual size of the calculated χ^2 is effected by the number of observed-expected comparisons, hence tabulated critical values get bigger as degrees of freedom increase, but for any constant degree of freedom, the larger the calculated value, the less similar are the observed and expected observations, and the less likely the null hypothesis of equivalence.

Notice that the values that make up a calculated χ^2 are divided by the sizes of the expected observations. This means that very small expected quantities may greatly increase a χ^2 value, even though the actual difference between observed and expected is relatively small. To avoid overinflation of the calculated χ^2 we must not allow expected values to become too small. Following accepted conventions, we will insist that all *expected* values must be greater than or equal to three. Note that there is no limit to the size of observed values (zero is fine). The second example (below) indicates one way of dealing with very small expected quantities without throwing away data.

EXAMPLES

a. Honeybees were allowed to forage at four different colors of artificial flowers. The number of bees at each flower 30 minutes after the onset of foraging was counted.

	FLOWER COLOR			
	BLUE	GREEN	RED	YELLOW
Number of bees	140	25	27	27

To test the null hypothesis that an equal number of bees visited each flower color, calculate the expected numbers of bees per flower assuming this null hypothesis were true, and compare these expected values with the observed values.

If an equal number visited all flower types, there would have been

$$(140 + 25 + 27 + 27)/4 = 54.75 \text{ bees/flower}$$

$$\chi^2 = \frac{(140 - 54.75)^2}{54.75} + \frac{(25 - 54.75)^2}{54.75} + \frac{(27 - 54.75)^2}{54.75} + \frac{(27 - 54.75)^2}{54.75}$$

$$= 177.0; \text{ degrees of freedom} = (4 - 1) = 3$$

Since our calculated value (177.0) is much larger than the tabulated χ^2 value for 3 degrees of freedom (7.8) and 0.05 significance level (Table 7), we reject the null hypothesis. We might continue the analyses by investigating the difference in the number of bees observed at green, red and yellow flowers. The expected value in this case would be (25 + 27 + 27) = 26.33. The χ^2 value would be 0.10 and would have 2 degrees of freedom.

b. Honeybees were observed foraging at flowers in a field in northern Virginia. The number of bees at each type of flower was recorded. The abundance of flowers of each type was then determined by counting all of the flowers in several sample transects mapped onto the field. Did the foraging bees use the flowers in proportion to their abundance?

FLOWER SPECIES	FLOWER ABUNDANCE	NUMBER OF BEES VISITING
Solidago sp.	692	79
Daucus carota	205	9
Aster pilosus	174	0
Bidens frondosa	151	0
Carduus discolor	53	1
Impatiens capensis	18	1

In this case, the null hypothesis is that bees foraged in direct proportion to flower abundance. If this is true, then we expect that the fraction of bees foraging at each type of flower should be equivalent to the fraction of flowers of that type in the field. For example, there were a total of 1293 flowers in the sample of flowers. Fifty-three of them (or 4.10%) were *C. discolor*; hence the proportion of *C. discolor* in the field was 0.0410. Since a total of 90 bees were observed foraging, we would expect (90)(0.0410) = 3.69 to be on *C. discolor* (rounding to whole bees gives 4). We can calculate other expected frequencies in a similar manner:

FLOWER SPECIES	OBSERVED BEES	EXPECTED BEES
Solidago sp.	79	48
D. carota	9	14
A. pilosus	0	12
B. frondosa	0	11
C. discolor	1	4
I. capensis	1	1

$\chi^2 = 46.61$; degrees of freedom = 5 — 1 = 4; probability level < 0.05.

As previously explained, expected values of less than 3 are not allowed when using the χ^2 test. Since the expected value for *I. capensis* was less than 3, it cannot be used. To avoid losing data, we eliminate the small expected value by combining it with *C. discolor* to form a new class which we can simply label "other." Thus the expected value for the class "other" is 5 and the observed value is 2. Using the class "other" we compared 5 categories; thus degrees of freedom are 4.

Since our calculated χ^2 value (46.9) is much greater than the tabulated χ^2 at 4 degrees of freedom and the 0.05 significance level (9.5; Table 7), we reject our null hypothesis and conclude that the bees were not foraging in proportion to flower abundance. Inspection of the observed numbers, and those expected for random foraging, shows that bees seemed to use *Solidago* more, and *B. frondosa*, *A. pilosus*, and *D. carota* much less, than expected by chance.

The Chi-Square Test: Contingency Tables

APPROPRIATE USES OF THE ANALYSIS

Suppose we have a set of observations that can be classified according to two types of different attributes. For example, suppose we have collected all of the salamanders in a patch of forest. The salamanders can be divided into different species, and they can be divided into groups according to the types of substrate they were captured on. We may be interested in knowing whether there is any association between species and substrate. We can evaluate the independence of various factors using contingency tables and the χ^2 statistic.

This type of analysis is appropriate whenever the sample can be divided into discrete categories, the probability distributions of which are unknown.

CALCULATIONS

a. Arrange the data in tabular form.

SECOND	FIRST FACTOR		
FACTOR	LEVEL A	LEVEL B	LEVEL C
Level 1	N_{a1}	N_{b1}	N_{c1}
Level 2	N_{a2}	N_{b2}	N_{c2}
Level 3	N_{a3}	N_{b3}	N_{c3}

1. Note that each factor comes in three levels in this example.
2. Note that each N refers to the number of individual observations in that cell of the table. For example, N_{b2} is the sample size for observations having the first variable in condition B and the second in condition 2.

b. Calculate the sums for each row and column of the table.

SECOND	FIRST FACTOR			
FACTOR	LEVEL A	LEVEL B	LEVEL C	SUMS
Level 1	N_{a1}	N_{b1}	N_{c1}	Sum 1
Level 2	N_{a2}	N_{b2}	N_{c2}	Sum 2
Level 3	N_{a3}	N_{b3}	N_{c3}	Sum 3
Sums	Sum A	Sum B	Sum C	Total

c. Calculate expected frequencies.

The probability of an event is equal to the frequency of that event divided by the frequency of all events under consideration. Thus the probability of the first factor occurring at level A equals the frequency of factor 1, level A (= Sum A) divided by the frequency of all factors and levels (= Total):

$$p \text{ (factor 1, level A)} = \text{Sum A/Total}$$

If two events are independent, then the probability of observing both events at the same time is equal to the probability of the first event times the probability of the second event. This simple rule allows us to calculate expected probabilities if we use independence as our null hypothesis.
If the null hypothesis is true, then the probability of observing the first factor in level A and the second factor in level 1 at the same time is equal to the probability of observing factor 1 in level A times the probability of observing factor 2 in level 1:

$$p \text{ (factor 1, level A)} = \text{Sum A/Total}$$

$$p \text{ (factor 2, level 1)} = \text{Sum 1/Total}$$

$$p \text{ (both conditions at once if independent)} = \text{(Sum A/Total)} \times \text{(Sum 1/Total)}$$

The expected frequency for an event is equal to the probability of that event times the total sample size, hence the expected frequency of observations at factor 1, level A and factor 2, level 1 is

$$\text{(Sum A/Total)} \times \text{(Sum 1/Total)} \times \text{Total} = \text{(Sum A} \times \text{Sum 1)/Total}$$

The other expected frequencies in the table are calculated in analogous fashion.

d. Compare observed and expected frequencies using the χ^2 statistic:

$$\chi^2 = \sum \frac{(\text{observed} - \text{expected})^2}{\text{expected}}$$

e. Determine degrees of freedom for the test.

$$\text{degrees of freedom} = (\text{no. of rows} - 1) \times (\text{no. of columns} - 1)$$

f. To evaluate the null hypothesis that there was no association, compare the calculated χ^2 with the tabular value for the appropriate degrees of freedom (Table 7). Reject the null if the calculated value exceeds the tabular value.

EXAMPLE

A single *Pipistrellus subflavus* bat was flown in an experimental chamber filled with obstacles. The obstacles in the chamber were thin nylon strings arranged in two types of patterns: 2-string arrangements and 3-string arrangements. The bat's behavior was measured in terms of the number of obstacles it hit and the number it missed during its flight. We want to know whether the performance of the bat (hits and misses) is independent of the arrangement of the obstacles (2-string and 3-string).

a. Collate the data.

		PERFORMANCE	
		HITS	MISSES
Obstacle	2-string	20	77
Pattern	3-string	19	88

b. Find row and column totals.

	HITS	MISSES	SUMS
2-string	20	77	97
3-string	19	88	107
Sums	39	165	204

c. Find expected values.

2-string hits (expected) = (97 x 39)/204 = 18.5
2-string misses (expected) = (97 x 165)/204 = 78.5
3-string misses (expected) = (107 x 165)/204 = 86.5
3-string hits (expected) = (107 x 39)/204 = 20.5

d. Calculate the χ^2 statistic.

$$\chi^2 = (20 - 18.5)^2/18.5 + (19 - 20.5)^2/20.5 + (77 - 78.5)^2/78.5 + (88 - 86.5)^2/86.5 = 0.286$$

The observed χ^2 in this example (0.03) is much smaller than the tabular value at 1 degree of freedom and the 0.05 significance level (3.8; Table 7), hence we do not have sufficient grounds to reject the null hypothesis, and we conclude that bat performance was not affected by the number of strings hanging in the bat's flight path.

The Chi-Square Test: Maximum Likelihood Comparisons

APPROPRIATE USES OF THE ANALYSIS

Suppose we have two separate frequency distributions of events, and wish to know whether they are both drawn form the same underlying distribution. (Put another way, we want to know if the two observed distributions differ significantly from each other). There is no extrinsic expected distribution in this case, but the contingency table analysis can be used to compare the two observed distributions with an expected distribution calculated directly from the observed values.

CALCULATIONS

a. Determine the expected frequency for each event in each distribution.

   Expected = $(N_i)(p)$, where N_i = sample size for that distribution
   p = probability of obtaining that event

   1. N_i is the total number of all observations in each of the observed distributions. There will thus be one N_i for each distribution.
   2. For each possible event, p is the number of times that event was observed in the first distribution, plus the number of times it was observed in the second distribution, all divided by the total number of all events in the two combined distributions.

b. Calculate the χ^2 statistic using the standard formula:

$$\chi^2 = \sum \frac{(\text{observed} - \text{expected})^2}{\text{expected}}$$

c. Calculate degrees of freedom associated with the test, where df = $K - 1$; K is the number of categories of events in the combined distribution.

d. Reject the null hypothesis of equivalence of both sets of frequencies if the calculated χ^2 exceeds the tabulated value for the appropriate degrees of freedom (Table 7).

EXAMPLE

We sampled mottled sculpins (*Cottus bairdi*) breeding at natural and at experimental spawning sites. The data reflect the number of mates breeding with individual males:

NUMBER OF FEMALES PER MALE	EXPERIMENTAL	NATURAL	TOTAL
1	26	13	39
2	23	12	35
3	17	9	26
4	11	7	18
5 or more	16	3	19
N	$N_1 = 93$	$N_2 = 44$	$N_1 + N_2 = 137$

Did the experimental male have the same distribution of females/males as did the natural males?

If the null hypothesis of equality is true, then the best estimate of the true, underlying distribution of mates/male is obtained by pooling both samples. For example, the overall probability (p) of a male having one mate would be

$$(26 + 13)/(93 + 44) = 39/137 = 0.2847$$

The expected number of males at experimental sites with one mate would then be

$$(0.2847)(93) = 26 \text{ (rounding off to whole females)}$$

The expected number of males occupying natural spawning sites with only one mate would be

$$(0.2847)(44) = 13$$

Similarly, other expected values can be determined:

NUMBER OF FEMALES/MALES	p	EXPERIMENTAL		NATURAL	
		OBS.	EXP.	OBS.	EXP.
1	.2847	26	26	13	13
2	.2555	23	24	12	11
3	.1898	17	18	9	8
4	.1314	11	12	7	6
5 or more	.1386	16	13	3	6
$N =$		93	93	44	44

All rounding errors have been accumulated in the last category in this example. Observed and expected columns must have the same total N values.

$$^2 = (26 - 26)^2/26 + (13 - 13)^2/13 \ldots + (3 - 6)^2/6$$

$$= 2.79$$

$$\text{df} = 5 - 1 = 4$$

Since our calculated χ^2 (2.8) is much smaller than the tabulated value (9.5; Table 7), we conclude that the two distributions are not different from each other in their shapes. Biologically, this means that we have no evidence that males at experimental sites attracted mates differently from those males at natural sites.

Analysis of Dispersion Patterns: The Poisson Distribution

APPROPRIATE USES OF THE ANALYSIS

There are many ways of analyzing dispersion patterns. Here we will examine one way that is appropriate whenever data can be collected on the number of individuals per sample. The exact nature of a sample may be determined by the biology of the animal or the design of the experiment. For example, we might count the number of aphids per leaf, mosquito larvae per puddle, or blister beetles per goldenrod stem. Other situations do not provide such convenient samples, however, and sample units must be imposed by the experimenter. For example, we might count the number of fiddler crabs per square meter of beach, the number of isopods resting beneath uniformly square boards, the number of cricket chirps per second, or the number of courtship behaviors per hour.

Once frequency data have been collected, they can be conveniently analyzed by comparing them to an expected distribution of frequencies with a chi-square test. Because a truly random distribution can be generated more easily than uniform or clumped patterns, the most commonly used expected distribution is a random one.

Deviation from randomness is either toward uniformity or toward clumping (see figure).

CALCULATIONS

a. Organize collected data as a frequency distribution showing the number of individuals per sample unit.

b. Calculate the mean and variance for the frequency distribution.

c. Determine the expected frequencies for each class in the distribution for a random distribution of that mean. Use either technique 1 or 2.

Technique 1. Relative expected frequencies for a random distribution of individuals per sample fit a Poisson distribution. The frequency of samples with i individuals per sample is equal to

$(e^{-\overline{Y}})(\overline{Y}^i)/i!$ where

e is the base of natural logs

$\overline{Y}$ is the mean

i is the number of individuals per sample

$i!$ means i factorial

For example, if the mean number of individuals per sample is 2, the relative frequency of samples having zero individuals would be

$(e^{-2})(2^0)/0! = 0.135$ (remembering that by definition, $0! = 1$)

Similarly, the number of samples with 1, 2, 3, etc. individuals can be determined.

Technique 2. Look up the relative expected frequencies for each number of individuals per sample in Table 8. Means are given at the top of each column in the table; frequencies are arranged down the sides.

d. Determine the expected number of observations in each class of the distribution by multiplying the expected frequencies obtained above by the sample size (total number of samples used).

e. Compare expected and observed values using a chi-square test, with the null hypothesis that your observed frequency distribution does not differ from randomness. The appropriate number of degrees of freedom is the number of categories compared minus two. Pool categories that have expected values of less than three. Reject the null hypothesis if the calculated χ^2 exceeds the tabulated value at the appropriate degrees of freedom.

If the null hypothesis is not rejected, conclude that the organisms were exemplifying a random dispersion pattern. If your null hypothesis is rejected, determine deviations from randomness by examining the coefficient of dispersion.

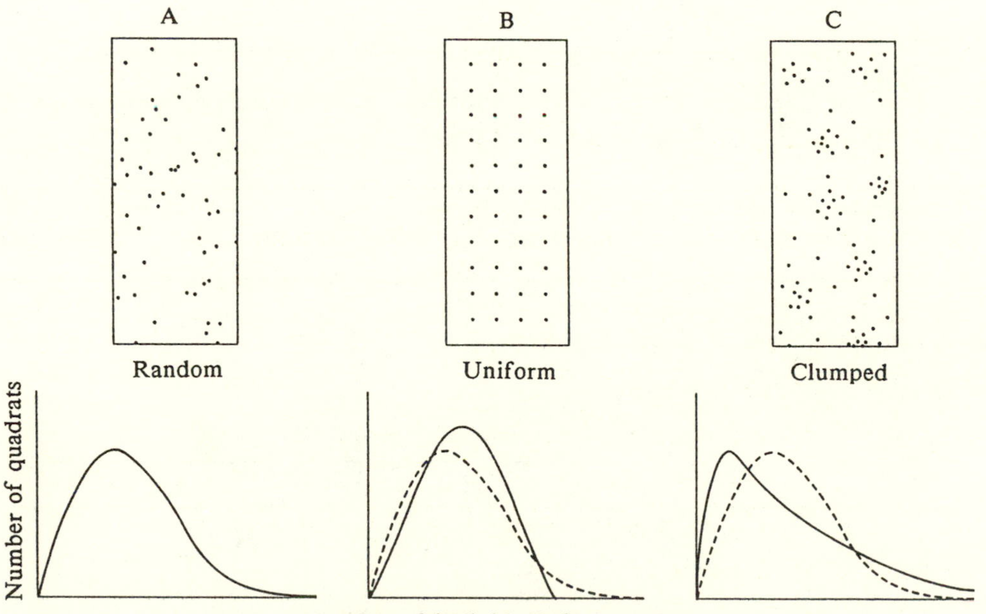

The three types of dispersion patterns. The lower figures graph the frequency distribution for number of individuals per quadrat (e.g., fiddler crabs per m² of beach); the plot of a random distribution is repeated as a dashed line in **(B)** *and* **(C)**.

1. **Coefficients of dispersion** are the ratio of the variance to the mean for a frequency distribution. A truly random (Poisson) distribution has a coefficient of dispersion of 1. Clumped patterns have values greater than 1, uniform patterns have values less than 1.
2. While there is no universally accepted test for the significance of a coefficient of dispersion, a test used for large samples is

$$\chi^2 = QS^2/\bar{Y} \text{ where } Q \text{ is the number of samples}$$

Significance of the χ^2 statistic is evaluated by looking at the appropriate table using $Q - 1$ degrees of freedom (Table 7). Note that this test is only valid if the mean is greater than 5 and Q is greater than 15; thus applications are very limited.

EXAMPLE

The following data represent the numbers of male and female sculpins (*Cottus bairdi*) resting beneath 12 cm × 12 cm pieces of slate in Anderson Creek, Champaign County, Ohio on April 19, 1975.

NUMBER OF INDIVIDUALS PER TILE	MALES	FEMALES
0	4	48
1	66	23
2	5	3
3	0	1
4	0	0

a. Treat each sex separately. The descriptive statistics, Poisson probabilities, and expected values for males are given below:

NUMBER OF MALES PER TILE	OBSERVED	PROBABILITY	EXPECTED
0	4	0.3642	27.32
1	66	0.3679	27.59
2	5	0.1858	13.94
3	0	0.0625	4.69
4	0	0.0196	1.46
	75	1.0000	75.00

$\bar{Y}$ = 76 males / 75 tiles = 1.01

s^2 = .12

χ^2 = 85.26

df = 2

CD = 0.12

Table 8. Expected frequencies for the Poisson distribution.

	MEAN											
CLASS	0.10	0.20	0.30	0.40	0.50	0.60	0.70	0.80	0.90	1.00	1.20	1.40
0	0.90	0.82	0.74	0.67	0.61	0.55	0.50	0.45	0.41	0.37	0.30	0.25
1	0.09	0.16	0.22	0.27	0.30	0.33	0.35	0.36	0.37	0.37	0.36	0.34
2	0.01*	0.02	0.03	0.05	0.08	0.10	0.12	0.14	0.16	0.18	0.22	0.24
3			0.01	0.01	0.01	0.02	0.03	0.04	0.05	0.06	0.09	0.11
4								0.01	0.01	0.02	0.03	0.04
5											0.10	0.01
6												0.01

	MEAN											
CLASS	1.60	1.80	2.00	2.20	2.40	2.60	2.80	3.00	3.20	3.40	3.60	3.80
0	0.20	0.16	0.14	0.11	0.09	0.07	0.06	0.05	0.04	0.03	0.03	0.02
1	0.32	0.30	0.27	0.24	0.22	0.19	0.17	0.15	0.13	0.11	0.10	0.08
2	0.26	0.27	0.27	0.27	0.26	0.25	0.24	0.22	0.21	0.19	0.18	0.16
3	0.14	0.16	0.18	0.20	0.21	0.22	0.22	0.22	0.22	0.22	0.21	0.20
4	0.06	0.07	0.09	0.11	0.12	0.14	0.16	0.17	0.18	0.19	0.19	0.19
5	0.02	0.03	0.04	0.05	0.06	0.07	0.09	0.10	0.11	0.13	0.14	0.15
6		0.01	0.01	0.02	0.02	0.03	0.04	0.05	0.06	0.07	0.08	0.09
7					0.02	0.03	0.02	0.02	0.03	0.03	0.04	0.05
8								0.01	0.01	0.01	0.02	0.02
9								0.01	0.01	0.02	0.01	0.04

(Modified from Rohlf, F.J. and R.R. Sokal. 1969. *Statistical Tables*, W.H. Freeman.)
*There are in fact an infinite number of classes; their frequencies, however, become vanishingly small. The final frequencies given in each column sum all succeeding frequencies.

b. Since the calculated χ^2 (85.3) greatly exceeds the tabulated value for 2 degrees of freedom (6.0) in Table 7, we conclude that the distribution of males is significantly different from a random distribution at the 0.05 probability level. Since the coefficient of dispersion is 0.12, we conclude that males were uniformly dispersed, and appeared to avoid one another: there was typically only one male per tile.

c. Proceed with the analysis for females in exactly the same manner. Notice that this analysis says nothing about interactions between the sexes, and deals only with interactions among like-sex individuals.

A Special Case: The Truncated Poisson

APPROPRIATE USES OF THE ANALYSIS

The analyses of dispersion presented so far have mandated a knowledge of the zero values of the random variable. For example, when we calculated the expected number of male sculpins per tile, we used the average number observed in the field. Tiles that were unoccupied by males were included in the calculation of the average. While we may be able to design experiments so that we can observe the zero class, we may occasionally be forced to deal with a data set in which the zero class in unknown. For example, suppose we simply moved through a stream sampling naturally-occurring rocks for sculpin nests. Sites with eggs would be easily counted, but what about rocks with no eggs? Including them in the zero class would be inaccurate, since some of them were probably not suitable as spawning sites. Consequently we can only count the nests with eggs, and we cannot know how many sites without eggs actually belong in the zero class. Data of this sort can be analyzed by calculating the expected frequencies for the truncated Poisson.

CALCULATIONS

In the conditional Poisson, the frequency of samples with i individuals per sample is:

$$e^{-\lambda}\ \lambda/i!(1 - e^{-\lambda}),\ \text{where } i = 1, 2, 3, ..., \ \lambda > 0$$

and

$$\lambda/(1 - e^{-\lambda}) = \overline{Y}$$

Where e is the base for natural logs
$\overline{Y}$ is the mean of the observed distribution
and i is the number of individuals per sample

A SHORTCUT

Although it is possible to solve for the expected frequencies using a hand calculator, it is more convenient to use Table 9 for the frequencies in a truncated Poisson.

Proceed with the chi-square calculations as you did with the normal Poisson analysis. Note that the coefficient of dispersion is inappropriate when using the truncated Poisson, since the mean is inflated by the missing zero class.

EXAMPLE

The following data represent the brood sizes for Chinese hamsters (*Cricetulus griseus*) raised in captivity.

NUMBER OF WEANLINGS	FREQUENCY
1	1
2	3
3	6
4	13
5	7
6	6
7	6
8	0
9	1

a. Mean brood size = Total offspring/Total broods
= 199/43 = 4.6

b. Calculate the expected values by rounding to the closest average in Table 9 (= 4.5).

NUMBER OF WEANLINGS	OBSERVED	EXPECTED
1	1	2.2
2	3	5.2
3	6	7.3
4	13	8.2
5	7	7.3
6	6	5.6
7	6	3.4
8	0	1.7
9	1	2.1

c. Calculate the value of χ^2 remembering that expected values cannot be less than 3, and therefore pooling broods of size 1 and 2 (i.e.,≤ 2) and 8 and 9 (i.e.,≥ 8).

$$\chi^2 = 8.70$$

$$df = 7 - 2 = 5$$

d. Compare the calculated χ^2 value (8.7) with that in Table 7 for 5 degrees of freedom. Since the calculated value is less than critical value of 11.1, we do not have sufficient reason to reject the null hypothesis that broods were randomly distributed, despite the fact that 8 of the 43 broods (20%) had at least 7 young.

Table 9. Expected frequencies for the truncated Poisson.

	MEAN								
CLASS	1.10	1.20	1.30	1.40	1.50	1.60	1.80	2.00	2.20
1	0.91	0.82	0.75	0.68	0.63	0.57	0.48	0.41	0.34
2	0.08	0.15	0.20	0.24	0.27	0.29	0.32	0.32	0.32
3	0.01*	0.02	0.04	0.06	0.08	0.10	0.14	0.17	0.20
4		0.01	0.01	0.02	0.02	0.02	0.05	0.07	0.09
						0.01	0.01	0.02	0.03
								0.01	0.01
									0.01

	MEAN							
CLASS	2.40	2.60	2.80	3.00	3.50	4.00	4.50	5.00
1	0.29	0.25	0.21	0.18	0.12	0.08	0.05	0.03
2	0.31	0.29	0.27	0.25	0.20	0.16	0.12	0.09
3	0.22	0.23	0.23	0.24	0.23	0.20	0.17	0.14
4	0.11	0.13	0.15	0.17	0.19	0.20	0.19	0.18
5	0.05	0.06	0.08	0.09	0.13	0.16	0.17	0.18
6	0.02	0.01	0.02	0.04	0.07	0.10	0.13	0.15
		0.02	0.01	0.02	0.04	0.06	0.08	0.10
			0.01	0.01	0.02	0.03	0.04	0.06
						0.01	0.02	0.04
							0.01	0.03

(Modified from Cohen, A.C. 1960. Estimating the Parameter in a Conditional Poisson Distribution. *Biometrics* [June], pp. 203-211.)
*As in Table 8, the final frequencies given in each column sum all succeeding frequencies.

Analysis of Binomial Distribution

APPROPRIATE USES OF THE ANALYSIS

Binomial distributions are composed of dichotomous data. Such data might record hits and misses, pluses and minuses, rights and lefts, males and females, accepts and rejects, or any other situation in which some data represent "successes" and the others represent "failures." Not all dichotomous data sets are binomial, and one of the things that we might want to know about a particular observed set is whether or not it is binomial. If it is, there is random assortment of the successes and failures. If it is not, there is some nonrandom pattern. This analysis is analogous to the use of a chi-square test to compare observed values to those expected using a Poisson distribution.

CALCULATIONS

a. Sort the entire data set into subsets of equivalent sample sizes. For example, if the data set represents the number of sons and daughters in families of 1-8 children, sort the families according to family size.

b. Determine the proportion of "successes" for each subset. Define "successes" so that this proportion is less than or equal to 0.50. Thus "successes" is either the proportion of sons or daughters, whichever is less.

c. Determine the relative expected frequencies for a binomial distribution either by calculation or inspection of Table 10.

   1. By calculation: The probability of observing X successes in N trials when the proportion of successes is equal to p is:

   $$\frac{N!}{X!(N-X)!} \cdot p^{x}(1-p)^{n-x}$$

   For example, the probability of observing 3 daughters in a family of 6 children would be 0.3125 if the probability of a daughter were 0.5.

   2. By inspection: Consult Table 10 for the relative expected frequencies. Subset size (N) is given in the first column, and is followed by the number of successes (X) in the second column. Various possible proportions of successes are arrayed across the top of the table. The body of the table gives the exact probabilities for each.

d. Determine the expected number of observations in each class of the distribution by multiplying the frequencies obtained in step c times the sample size for the subset.

Table 10. Relative proportions for individual terms of the binomial distribution.

TRIALS OF SIZE (*n*)	NUMBER OF SUCCESSES (*x*)	SELECTED PROPORTIONS OF "SUCCESSES"									
		0.05	0.10	0.15	0.20	0.25	0.30	0.35	0.40	0.45	0.50
2	0	0.90	0.81	0.72	0.64	0.56	0.49	0.42	0.36	0.30	0.25
	1	0.10	0.18	0.26	0.32	0.38	0.42	0.46	0.48	0.50	0.50
	2		0.01	0.02	0.04	0.06	0.09	0.12	0.16	0.20	0.25
3	0	0.86	0.73	0.61	0.51	0.42	0.34	0.27	0.22	0.17	0.12
	1	0.14	0.24	0.33	0.38	0.42	0.44	0.44	0.43	0.41	0.38
	2		0.03	0.06	0.10	0.14	0.19	0.24	0.29	0.33	0.38
	3				0.01	0.02	0.03	0.05	0.06	0.09	0.12
4	0	0.81	0.66	0.52	0.41	0.32	0.24	0.18	0.13	0.09	0.06
	1	0.17	0.29	0.37	0.41	0.42	0.41	0.38	0.35	0.30	0.25
	2	0.02	0.05	0.10	0.15	0.21	0.26	0.31	0.35	0.37	0.38
	3			0.01	0.02	0.04	0.08	0.11	0.15	0.20	0.25
	4					0.01	0.01	0.02	0.02	0.04	0.06
5	0	0.77	0.59	0.44	0.33	0.24	0.17	0.12	0.08	0.05	0.03
	1	0.20	0.33	0.39	0.41	0.40	0.36	0.31	0.26	0.21	0.16
	2	0.03	0.07	0.14	0.20	0.26	0.31	0.34	0.35	0.34	0.31
	3		0.01	0.03	0.05	0.09	0.13	0.18	0.32	0.28	0.31
	4				0.01	0.01	0.03	0.05	0.08	0.11	0.16
	5								0.01	0.01	0.03
6	0	0.74	0.53	0.38	0.26	0.18	0.12	0.08	0.05	0.03	0.02
	1	0.23	0.35	0.40	0.40	0.36	0.30	0.24	0.19	0.14	0.09
	2	0.03	0.10	0.18	0.25	0.30	0.32	0.33	0.31	0.28	0.23
	3		0.02	0.04	0.08	0.13	0.18	0.23	0.28	0.30	0.31
	4				0.01	0.03	0.06	0.10	0.14	0.19	0.23
	5						0.02	0.02	0.03	0.06	0.09
	6										0.03
7	0	0.69	0.48	0.32	0.21	0.13	0.08	0.05	0.03	0.02	0.01
	1	0.26	0.37	0.40	0.37	0.31	0.25	0.18	0.13	0.09	0.05
	2	0.04	0.12	0.21	0.28	0.31	0.32	0.30	0.26	0.21	0.16
	3	0.01	0.02	0.06	0.11	0.17	0.22	0.27	0.29	0.29	0.27
	4		0.01	0.01	0.03	0.06	0.10	0.14	0.19	0.24	0.27
	5					0.02	0.03	0.05	0.08	0.11	0.16
	6							0.01	0.02	0.03	0.05
	7									0.01	0.02
	9										0.01
8	0	0.66	0.43	0.27	0.17	0.10	0.06	0.03	0.02	0.01	0.00
	1	0.28	0.38	0.39	0.34	0.27	0.20	0.14	0.09	0.05	0.03
	2	0.05	0.15	0.24	0.29	0.31	0.30	0.26	0.21	0.16	0.11
	3	0.01	0.03	0.08	0.15	0.21	0.25	0.28	0.28	0.26	0.22
	4		0.01	0.02	0.05	0.09	0.13	0.17	0.23	0.26	0.28
	5				0.01	0.02	0.05	0.08	0.12	0.17	0.22
	6						0.01	0.02	0.04	0.07	0.11
	7								0.01	0.02	0.03
	8										

(continued)

Table 10. (Continued)

TRIALS OF SIZE (*n*)	NUMBER OF SUCCESSES (*x*)	SELECTED PROPORTIONS OF "SUCCESSES"									
		0.05	0.10	0.15	0.20	0.25	0.30	0.35	0.40	0.45	0.50
9	0	0.63	0.39	0.23	0.13	0.08	0.04	0.02	0.01	0.01	0.00
	1	0.30	0.39	0.37	0.30	0.22	0.16	0.10	0.06	0.03	0.02
	2	0.06	0.17	0.26	0.30	0.27	0.26	0.22	0.16	0.11	0.07
	3	0.01	0.04	0.11	0.18	0.23	0.27	0.27	0.25	0.21	0.16
	4		0.01	0.03	0.07	0.12	0.17	0.22	0.25	0.26	0.25
	5				0.02	0.04	0.07	0.12	0.17	0.21	0.25
	6					0.01	0.02	0.04	0.07	0.12	0.16
	7							0.01	0.02	0.04	0.07
	8								0.01	0.01	0.02
	9										
10	0	0.60	0.35	0.19	0.11	0.06	0.03	0.01	0.01	0.00	0.00
	1	0.32	0.39	0.35	0.26	0.19	0.12	0.07	0.04	0.02	0.01
	2	0.07	0.19	0.28	0.30	0.28	0.23	0.18	0.12	0.08	0.04
	3	0.01	0.06	0.13	0.20	0.25	0.27	0.25	0.27	0.17	0.12
	4		0.01	0.04	0.09	0.15	0.20	0.24	0.25	0.24	0.21
	5			0.01	0.03	0.06	0.10	0.16	0.20	0.23	0.25
	6				0.01	0.01	0.04	0.07	0.11	0.16	0.20
	7						0.01	0.02	0.04	0.07	0.12
	8								0.01	0.03	0.04
	9										0.01

(Modified from Selby, S. M. (ed.). 1970. *Standard Mathematical Tables*. The Chemical Rubber Co., Cleveland.)

e. Compare the expected and observed values using a chi-square test and the null hypothesis that your observed frequency distribution does not differ from the binomial. The appropriate number of degrees of freedom is the number of categories compared minus two. Pool categories with expected values of less than three. Reject the null hypothesis if the calculated X^2 exceeds the tabulated value at the appropriate degrees of freedom (Table 7).

   If the null hypothesis is not rejected, conclude that the organisms were distributed in binomial fashion. This means that there was some constant probability (p) of successes (having some biological meaning that depends on that particular case being studied), and that the occurrence of a success was a random event.

   If the null hypothesis is rejected, determine how the observed distribution deviates from the binomial by inspecting observed and expected values. If there were excessive clumps of successes (e.g., many more families of 6 children with 6 daughters) or failures (lots of families with 0 daughters), the observed distribution is clumped. If there were too few clumps of successes (e.g., all families of 6 had 3 daughters and 3 sons), the observed distribution is uniform or repulsed.

EXAMPLE

The following data set gives the number of male and female sculpins (*Cottus bairdi*) found resting beneath stones in a creek in central Ohio. Suitable stones without fish could not be distinguished from unsuitable (uninhabitable) stones; hence there is no record for stones without any fish.

NUMBER OF MALES	NUMBER OF FEMALES				
	0	1	2	3	4
0	?	6	0	0	0
1	534	198	24	6	0
2	48	18	12	0	0
3	6	0	0	0	0
4	0	6	0	0	0

a. Divide the total data set into subsets representing groups having equal sizes. In this example, we will pay attention only to the subset of groups having 2 members.

NUMBER OF MALES	NUMBER OF FEMALES		
	0	1	2
0			0
1		198	
2	48		

b. Determine the probability of a success.

Number of females = 198

Number of males = 294

Total number of fish = 492

Proportion of females = 198/492 = 0.4

Proportion of males = 294/492 = 0.6

Therefore, the proportion of successes is defined as the proportion of females, or 0.4.

c. Determine the expected frequencies for a binomial distribution having a probability of success equal to 0.6 (Table 10). The expected frequencies are:

NUMBER OF FEMALES	FREQUENCY
0	0.36
1	0.48
2	0.16

d. Determine the expected numbers of groups by multiplying the frequencies by the sample size for groups of size 2:

NUMBER OF FEMALES	FREQUENCY	EXPECTED NUMBER
0	0.36	0.36 x 246 = 89
1	0.48	0.48 x 246 = 118
2	0.16	0.16 x 246 = 39

e. Compare the observed values with those expected using a χ^2 statistic. In this case, the calculated χ^2 value is 112, much larger than the critical value of 3.8 with 1 degree of freedom (Table 7). Inspection of the observed and expected values indicates far fewer groups having as an original only one female. This suggests that there is not a constant probability of finding a female, and that it is much more likely to find one female with a male than predicted by the binomial.

Multidimensional Tables: G *Tests*

APPROPRIATE USES OF THE ANALYSIS

Sometimes we collect data on more than one characteristic (factor) of an individual and want to know whether these factors are associated with one another. The chi-square contingency table analysis presented earlier allows us to evaluate association between two factors. This analysis can be expanded to include more than two factors. Unfortunately, the math involved for multidimensional chi-square contingency tables is difficult. Fortunately, the math involved in an analogous approach using the G statistic is much more manageable. Here we present an analysis designed to deal with the case of three factors, each having at least two levels. For example, we might have counted the number of individuals in each of four species of salamander found on each of two substrates types (e.g., wet, dry) at each of two times (day, night). We want to know whether there is any association between these factors, and if there is association, what its nature is. The null hypotheses we will be testing first is that there is no association between the three factors. If this is rejected, then we will continue to test subhypotheses to reveal interaction (i.e., association) between pairs of factors.

CALCULATIONS

Consider an experiment in which we have established three treatments, A, B and C. Consider further that each treatment has two states: Treatment A has states L and M; treatment B has states R and S; and treatment C has states X and Y. For example, in the egg lab we placed eggs of different sizes and colors in different habitats. We used two habitats, and two sizes and two colors of eggs. Habitat, size and color are the treatments. Habitat was broken into two states: forest and field; size was large or small; and color was either white or brown. Sometime after the start of the experiment we counted the number of surviving of eggs in each class. The counts are the entries in the table (n_1). In general terms we can array the data as follows:

		TREATMENT A	
TREATMENT B	TREATMENT C	STATE L	STATE M
State R	State X	n_1	n_5
	State Y	n_2	n_6
State S	State X	n_3	n_7
	State Y	n_4	n_8

A. Calculate the various marginal sums:

Treatment A: State L (ΣL) $= n_1 + n_2 + n_3 + n_4$
State M (ΣM) $= n_5 + n_6 + n_7 + n_8$

Treatment B: State R (ΣR) $= n_1 + n_2 + n_5 + n_6$
State S (ΣS) $= n_3 + n_4 + n_7 + n_8$

Treatment C: State X (ΣX) $= n_1 + n_3 + n_5 + n_7$
State Y (ΣY) $= n_2 + n_4 + n_6 + n_8$

Grand Total N: $N_t = n_1 + n_2 + n_3 + n_4 + n_5 + n_6 + n_7 + n_8$

B. Define a suite of sums that will be used to calculate interactions among the treatments:

For interaction between

A and B: $LR = n_1 + n_2$ $LS = n_3 + n_4$
$MR = n_5 + n_6$ $MS = n_7 + n_8$

A and C: $LX = n_1 + n_3$ $LY = n_2 + n_4$
$MX = n_5 + n_7$ $MY = n_6 + n_8$

B and C: $RX = n_1 + n_5$ $RY = n_2 + n_6$
$SX = n_3 + n_7$ $SY = n_4 + n_8$

C. Calculation of a G statistic requires conversion of the data to the product of the number (say F) times the natural logarithm (ln) of that number. That is, code each term as $F \ln F$. Calculate the following 8 coded sums:

$$T = n_1 \ln n_1 + n_2 \ln n_2 + n_3 \ln n_3 + n_4 \ln n_4 + n_5 \ln n_5 + n_6 \ln n_6 + n_7 \ln n_7 + n_8 \ln n_8$$

$A = \Sigma L \ln \Sigma L + \Sigma M \ln \Sigma M$
$B = \Sigma R \ln \Sigma R + \Sigma S \ln \Sigma S$
$C = \Sigma X \ln \Sigma X + \Sigma Y \ln \Sigma Y$

The interaction between

A and B $= LR \ln LR + LS \ln LS + MR \ln MR + MS \ln MS$
A and C $= LS \ln LX + LY \ln LY + MX \ln MX + MY \ln MY$
B and C $= RX \ln EX + RY \ln RY + SX \ln SX + SY \ln SY$

$N = N_t \ln N_t$

D. There are several null hypotheses (H_0) we can examine with these data:

1. Treatments A, B, and C are independent. Rejection of this hypothesis indicates that one or more of the following may also be rejected.
2. There is no interaction between A and B.
3. There is no interaction between A and C.
4. There is no interaction between B and C.
5. There is no interaction among A, B, and C jointly.

The calculations for each of these null hypotheses are as follows:

1. A, B, and C are independent. Calculate the following G statistic

$$G_t = 2[(T + 2N) - (A + B + C)]$$

To test for significance compare the calculated G value to a χ^2 value (Table 7) with $abc - (a - 1) - (b - 1) - (c - 1) - 1$ degrees of freedom, where a, b and c are number of classes in treatments A, B and C respectively. In the case we have been discussing, the degrees of freedom

$$= (2)(2)(2) - (2 - 1) - (2 - 1) - (2 - 1) - 1 = 4$$

2. There is no interaction between A and B. Calculate the following G statistic:

$$G_{ab} = 2[(A \times B + N) - (A + B)]$$

This G value will have $ab - (a - 1) - (b - 1) - 1$ degrees of freedom.

3. There is no interaction between A and C. Calculate the following G statistic:

$$G_{ac} = 2[(A \times C + N) - (A + C)]$$

This G value will have $ab - (a - 1) - (b - 1) - 1$ degrees of freedom.

4. There is no interaction between habitat and color. Calculate the following G statistic:

$$G_{bc} = 2[(B \times C + N) - (B + C)]$$

This G value will have $bc - (b - 1) - (c - 1) - 1$ degrees of freedom.

5. There are no interactions among A, B and C jointly. Calculate the following G statistic:

$$G_{abc} = 2[(T + A + B + C) - (A \times B + A \times C + B \times C + N)]$$

This G value will have $(a - 1) + (b - 1) + (c - 1)$ degrees of freedom (1 df in this case).

Note that in each of the interaction tests (2-4), the G statistic is the difference between the interaction ($A \times B$, $A \times C$ or $B \times C$) and the sum of the single effects (for example $A + B$ in test 2). Since there is no term $A \times B \times C$, we calculate G_{abc} by subtraction since it is true that:

$$G_T = G_{ab} + G_{ac} + G_{bc} + G_{abc}$$

and it is also true that the degrees of freedom must have been partitioned so that they sum to the total degrees of freedom ($4 = 1 + 1 + 1 + 1$, in this example). Finally, if there is more than one degree of freedom in any class (more than 2 states) we can partition that class to look for the effects at a finer level.

Suppose that Treatment C had three states: X, Y and Z and that G_{ac} was significant. That result could be due to effects of A with X, Y or Z. Since G_{ac} will have $ac - (a - 1) - (b - 1) - 1$ degrees of freedom (df) or $(2 \times 3) - (2 - 1) - (3 - 1) - 1 = 2$ df, we can only perform two tests. However, if the two we perform are not significant, the third must be. These can be performed as simple 2 by 2 contingency tables.

When we calculate G_T, we may find that the variables we are studying are not independent. The subsequent tests we outlined provide a means of examining those possible effects that cause us to reject the null hypothesis. In examining those effects, we have partitioned the original G_T into components and tested to see which ones were significant. A little reflection will lead you to conclude that we exhausted all possible effects or combinations of effects and hence we should have accounted for all of the original variance in G_T.

E. In some cases, the sum of the two-way interaction terms may be greater than G_T. If that is the case, then the three-way interaction term (G_{abc}) will be negative. When that occurs it is necessary to repartition the interaction terms in the following manner. We will use lower case symbols for the new sums that we will need.

For interactions between

a and b:

$$lr = [(LX)(RX)/ \Sigma X] + [(LY)(RY)/ \Sigma Y]$$
$$ls = [(LX)(SX)/ \Sigma X] + [(LY)(SY)/ \Sigma Y]$$

$$mr = [(MX)(RX)/ \Sigma X] + [(MY)(RY)/ \Sigma Y]$$
$$ms = [(MX)(SX)/ \Sigma X] + [(MY)(SY)/ \Sigma Y]$$

a and c:

$$lx = [(LR)(RX)/ \Sigma R] + [(LS)(SX)/ \Sigma S]$$
$$ly = [(LR)(RY)/ \Sigma R] + [(LS)(SY)/ \Sigma S]$$

$$mx = [(MR)(RX)/ \Sigma R] + [(MS)(SX)/ \Sigma S]$$
$$my = [(MR)(RY)/ \Sigma R] + [(MS)(SY)/ \Sigma S]$$

b and c:

$$rx = [(LR)(LX)/ \Sigma L] + [(MR)(MX)/ \Sigma M]$$
$$ry = [(LR)(LY)/ \Sigma L] + [(MR)(MY)/ \Sigma M]$$

$$sx = [(LS)(LX)/ \Sigma L] + [(MS)(MX)/ \Sigma M]$$
$$sy = [(LS)(LY)/ \Sigma L] + [(MS)(MY)/ \Sigma M]$$

When the three-way interaction term is negative it implies that one or more of our two sums has been inflated by its interaction with the third variable. Each of the new sums is adjusted for that possibility by taking into account the relationship of each of the two terms in the interaction to the third term. These new values are subject to the restriction that the sum of the new interaction elements equals the sum of the original elements. For example: $L = lr + ls$.

The new coded sums are:

$$a \times b = LR \ln lr + LS \ln ls + MR \ln mr + MS \ln ms$$
$$a \times c = LX \ln lx + LY \ln ly + MX \ln mx + MY \ln my$$
$$b \times c = RX \ln rx + RY \ln ry + SX \ln sx + SY \ln sy$$

and we calculate the needed G values as before except that we substitute the new coded sums for their original counterparts.

$$G_{ab} = 2[(a \times b + N) - (A + B)]$$
$$G_{ac} = 2[(a \times c + N) - (A + C)]$$
$$G_{bc} = 2[(b \times c + N) - (B + C)]$$
$$G_{abc} = 2[(T + A + B + C) - (a \times b + a \times c + b \times c + N)]$$

EXAMPLE

Consider the following table of the distribution of eggs of different colors and color patterns among birds with different nesting habits.

		NEST LOCATION (L)	
COLOR (C)	PATTERN (P)	GROUND (G)	OTHER (O)
White (W)	Unmarked (U)	49	141
	Speckled (S)	13	134
Brown (B)	Unmarked (U)	22	4
	Speckled (S)	83	20

A. Calculate the marginal sums:

L: Location: $\Sigma G = 49 + 103 + 22 + 83 = 257$
$\Sigma O = 141 + 134 + 4 + 20 = 299$

C: Color: $\Sigma W = 49 + 103 + 141 + 134 = 427$
$\Sigma B = 22 + 83 + 4 + 20 = 129$

P: Pattern: $\Sigma U = 49 + 141 + 22 + 4 = 216$
$\Sigma S = 103 + 134 + 83 + 20 = 340$

N: Total sum: $N_t = 49 + 141 + 103 + 134 + 22 + 4 + 83 + 20 = 556$

B. Calculate the interactions between:

L and C: (Location and Color)
$GW = 49 + 103 = 152$
$GB = 22 + 83 = 105$
$OW = 141 + 134 = 275$
$OB = 4 + 20 = 24$

L and P: (Location and Pattern)
$GU = 49 + 22 = 71$
$GS = 103 + 83 = 186$
$OU = 141 + 4 = 145$
$OS = 134 + 20 = 154$

C and P: (Color and Pattern)
$WU = 49 + 141 = 190$
$WS = 103 + 134 = 237$
$BU = 22 + 4 = 26$
$BS = 83 + 20 = 103$

C. Transform to $N \ln N$ for calculation of appropriate G statistics

$$T = 49 \ln 49 + 141 \ln 141 + 103 \ln 103 + 123 \ln 123 + 22 \ln 22 + 4 \ln 4 + 83 \ln 83 + 20 \ln 20$$
$$= 191 + 698 + 477 + 656 + 68 + 6 + 367 + 60 = \mathbf{2523}$$

$$L = 257 \ln 257 + 299 \ln 299 = 1426 + 1704 = \mathbf{3130}$$
$$C = 427 \ln 427 + 129 \ln 129 = 2586 + 627 = \mathbf{3213}$$
$$P = 216 \ln 216 + 344 \ln 344 = 1161 + 2009 = \mathbf{3170}$$

$$L \times C = 152 \ln 152 + 105 \ln 105 + 275 \ln 275 + 24 \ln 24$$
$$= 764 + 489 + 1545 + 76 = \mathbf{2874}$$
$$L \times P = 71 \ln 71 + 186 \ln 186 + 145 \ln 145 + 154 \ln 154$$
$$= 303 + 972 + 722 + 776 = \mathbf{2773}$$
$$C \times P = 190 \ln 190 + 237 \ln 237 + 26 \ln 26 + 103 \ln 103$$
$$= 997 + 1296 + 85 + 477 = \mathbf{2855}$$
$$N = 560 \ln 560 = \mathbf{3544}$$

D. Calculate G for each of the following null hypotheses:

1. Location, color and pattern are independent:

$$G_T = 2[(T + 2N) - (L + C + P)]$$
$$= 2[(2523 + 2(3544)) - (3130 + 3213 + 3170)]$$
$= \mathbf{196}$ with $lcp - (l - 1) - (c - 1) - (p - 1) - 1$ degrees of freedom
$= \mathbf{4\ df}$

Comparison with the appropriate value in Table 7 shows that this is a significant χ^2. Location, color and pattern are not independent.

2. There is no interaction between location and color:

$$G_{lc} = 2[(L \times C + N) - (L + C)]$$
$$= 2[(2874 + 3544) - (3130 + 3213)]$$
$= \mathbf{150}$, with $lc - (l - 1) - (c - 1) - 1$ degrees of freedom $= \mathbf{1\ df}$

Again, a significant χ^2. There is a significant interaction between location and color.

3. There is no interaction between location and pattern:

$$G_{lp} = 2[(L \times P + N) - (L + P)]$$
$$= 2[(2773 + 3544) - (3130 + 3170)]$$
$= \mathbf{34}$, again with 1 df, and again a significant χ^2

There is a significant interaction between location and pattern.

4. There is no interaction between pattern and color:

$$G_{cp} = 2[(C \times P + N) - (C + P)]$$
$$= 2[(2855 + 3544) - (3213 + 3170)]$$
$= \mathbf{32}$, a significant χ^2 with 1 df

5. There is no significant joint interaction among location, color and pattern:

$$G_{lcp} = 2[(T + L + C + P) - (L \times C + L \times P + C \times P + N)]$$
$$= 2[2523 - (2874 + 2773 + 2855) + (3130 + 3213 + 3170) - 3544]$$
$$= \mathbf{-20} \text{ with 1 degrees of freedom}$$

Since G_{lcp} is negative we will have to recalculate the two interaction terms. Before doing so we will check our calculations indirectly by summing the various interaction G values to see if they sum to G_T.

$$150 + 34 + 32 - 20 = 196 = G_T$$

E. Calculations when the three-way interaction is negative:

1. Calculate the new cross products for interaction between:

l and *c*:

$$gw = [(GU)(WU)/U] + [(GS)(WS)/S]$$
$$= [(71)(190)/216] + [(186)(237)/340] = 192$$

$$gb = [(GU)(BU)/U] + [(GS)(BS)/S]$$
$$= [(71)(26)/216] + [(186)(103)/340] = 65$$

$$ow = [(OU)(WU)/U] + [(OS)(WS)/S]$$
$$= [(145)(190)/216] + [(154)(237)/340] = 235$$

$$ob = [(OU)(BU)/U] + [(OS)(BS)/S]$$
$$= [(145)(26)/216] + [(154)(103)/340] = 64$$

l and *p*:

$$gu = [(GW)(WU)/W] + [(GB)(BU)/B]$$
$$= [(152)(190)/427] + [(105)(26)/129] = 89$$

$$gs = [(GW)(WS)/W] + [(GB)(BS)/B]$$
$$= [(152)(237)/427] + [(105)(103)/129] = 168$$

$$ou = [(OW)(WU)/W] + [(OB)(BU)/B]$$
$$= [(275)(190)/427] + [(24)(26)/129] = 127$$

$$os = [(OW)(WS)/W] + [(OB)(BS)/B]$$
$$= [(275)(237)/427] + [(24)(103)/129] = 172$$

c and *p*:

$$wu = [(GW)(GU)/G] + [(OW)(OU)/O]$$
$$= [(152)(71)/257] + [(275)(145)/299] = 175$$

$$ws = [(GW)(GS)/G] + [(OW)(OS)/O]$$
$$= [(152)(186)/257] + [(275)(154)/299] = 252$$

$$bu = [(GB)(GU)/G] + [(OB)(OU)/O]$$
$$= [(105)(71)/257] + [(24)(145)/299] = 41$$

$$bs = [(GB)(GS)/G] + [(OB)(OS)/O]$$
$$= [(105)(186)/257] + [(24)(154)/299] = 88$$

Since the new cross products must sum to the old cross products, for example: $gu + gs = G$, we really only need to do half of the calculations listed above (provided you do the right ones.)

2. Calculate the new coded interaction terms:

$$
\begin{aligned}
l \times c &= GW \ln gw + GN \ln gb + OW \ln ow + OB \ln ob \\
&= 152 \ln 192 + 105 \ln 65 + 275 \ln 235 + 24 \ln 64 \\
&= 799 + 438 + 1501 + 100 \\
&= \mathbf{2838}
\end{aligned}
$$

$$
\begin{aligned}
l \times p &= GU \ln gu + GS \ln gs + OU \ln ou + OS \ln os \\
&= 71 \ln 89 + 186 \ln 168 + 145 \ln 127 + 154 \ln 172 \\
&= 319 + 953 + 702 + 793 \\
&= \mathbf{2767}
\end{aligned}
$$

$$
\begin{aligned}
c \times p &= WU \ln wu + WS \ln ws + BU \ln bu + BS \ln bs \\
&= 190 \ln 175 + 237 \ln 252 + 261 \ln 41 + 103 \ln 88 \\
&= 981 + 1310 + 97 + 461 \\
&= \mathbf{2849}
\end{aligned}
$$

3. Calculate G for each interaction term:

$$
\begin{aligned}
G_{\mathrm{lc}} &= 2[(l \times c + N) - (L + C)] \\
&= 2[(2838 + 3544) - (3130 + 3213)] \\
&= \mathbf{78}
\end{aligned}
$$

$$
\begin{aligned}
G_{\mathrm{lp}} &= 2[(l \times p + N) - (L + P)] \\
&= 2[(2767 + 3544) - (3130 + 3170)] \\
&= \mathbf{22}
\end{aligned}
$$

$$
\begin{aligned}
G_{\mathrm{cp}} &= 2[(c \times p + \mathrm{N}) - (C + P)] \\
&= 2[(2849 + 3544) - (3213 + 3170)] \\
&= \mathbf{20}
\end{aligned}
$$

$$
\begin{aligned}
G_{\mathrm{lcp}} &= 2[(T + L + C + P) - (l \times c + l \times p + c \times p + N)] \\
&= 2[(2523 + 3130 + 3213 + 3170) - (2838 + 2767 + 2849 + 3544)] \\
&= \mathbf{76}
\end{aligned}
$$

We can check our results by seeing if the G values sum to G_{T}:

$$78 + 22 + 20 + 76 = 196 = G_{\mathrm{T}}$$

They do.

Each of the G values has 1 df, and all are highly significant (Table 7). The bulk of the variability is accounted for by the interaction between location, color and pattern. Together they account for nearly 90% of G_{T}.

Runs Tests

APPROPRIATE USES OF THE ANALYSIS

We are often interested in the pattern of events as well as the final outcome. The runs test is designed to tell us something about the sequence of events. For example, we might flip a coin 10 times, and find that there were 5 heads and 5 tails. If we tested this outcome with a chi-square, we would conclude that the coin was "honest." If the sequence of heads and tails had been

HHHHH TTTTT

then we might doubt the results of the chi-square test. On the other hand, if the sequence were:

H T H T H T H T H T

then we might be equally skeptical about the honesty of the coin.

If we define a run as a sequence of identical events, which begins with a switch to that event and ends with a switch back to the other event, then the first example has too few runs (only 2), and the second has too many (10). The runs test is a way to decide whether there are too many or too few runs.

The runs test is only appropriate for discrete classes.

CALCULATIONS

a. Determine the number of runs in the sample.

b. Calculate the mean number of runs, where

$$\text{Mean number of runs} = \bar{R} = 1 + [(2N_1)(N_2)/(N_1 + N_2)]$$

c. Calculate the standard deviation (Sr) for the number of runs.

$$\text{Sr} = \frac{(2N_1)(N_2)[(2N_1)(N_2) - N_1 - N_2]}{(N_1 + N_2)^2\,(N_1 + N_2 - 1)}$$

Where N_1 and N_2 are the numbers of the first condition (e.g., heads) and the second condition (e.g., tails) in the sample.

d. Calculate t, where

$$t = \frac{|R - \bar{R}|}{\text{Sr}}$$

e. If the calculated t is greater than 1.96, then reject the null hypothesis.

RATIONALE

The runs test simply computes a t statistic to compare the number of observed runs with the mean number expected for a random sequence of events. Note that comprehensive statistics texts provide tables of critical values for the number of runs from small samples of different sizes. We have presented a version of this test suitable for larger samples (ideally $N_1 + N_2 = 40$). We further note that the test can be used for more than two states, since we can always separate them into the class we are interested in and a second class (i.e., all others).

EXAMPLE

When an adult guppy *(Poecilia reticulata)* was placed in an aquarium containing equal numbers of large (adult) and small (larval) *Daphnia*, it ate the prey in this order:

LLLLLL S L S LLLLLLLLLL
S LLLLLLLLL S LLLLLLLL
S LLL S L

a. Determine the number of runs.

$$R = 13$$

b. Calculate the mean.

$$R = 1 + [(2 \times 40 \times 6)/(40 + 6)] = 11.43$$

c. Calculate the standard deviation.

$$Sr = \frac{(2 \times 40 \times 6)(2 \times 40 \times 6 - 40 - 6)}{(40 + 6)^2(40 + 6 - 1)}$$

$$= 2.19$$

d. Calculate t.

$$t = \frac{13 - 11.43}{2.19}$$

$$= .72$$

e. Compare the calculated t (0.72) with the critical value of 1.96. Since the calculated t is smaller, we cannot reject the null hypothesis that prey were eaten in random order, even though more large prey were actually consumed.

Circular Statistics: The Rayleigh Test

APPROPRIATE USES OF THE ANALYSIS

When we collect data on orientation, or on daily activity patterns, we are dealing with data arrayed along an ordinate that circles back on itself. For example, as we move around a compass (starting at true north), we first face away from north, and then once we have moved through true south, we begin to face back toward north. If we only looked at the degree bearings, we would conclude that a reading of 1° and one of 359° were a long way apart, yet they are only one degree on either side of true north.

In order to deal with data of this sort, we will need to transform it so that the effects of circularity are accounted for. Once this is done, we need to know whether observations cluster around a mean or are random. The Rayleigh test tells us whether observed departure bearings or daily activity cycles are random, and allows calculations of mean times or bearings.

CALCULATIONS

a. Determine the sine and cosine of each angle (Table 11).

b. Calculate the sum of the sines (V), and the sum of the cosines (W).

c. Calculate R.

$$R = \sqrt{V^2 + W^2}$$

d. Calculate the mean sine, mean cosine, and r.

$$\overline{\sin} = V/N$$

$$\overline{\cos} = W/N$$

$$r = R/N$$

e. Calculate the sine and cosine of the mean angle.

$$\sin = \overline{\sin}/r$$

$$\cos = \overline{\cos}/r$$

f. Calculate w, where

$$w = R^2/N$$

g. If $w > 3.00$, reject the null hypothesis; w does not change greatly with sample size: $p < 0.05$ when w is > 3.00 for all sample sizes.

RATIONALE

To unfold the distribution we convert each observation to its sine and cosine. The sine of an angle in a right triangle is the side opposite divided by the hypotenuse, and the cosine of an angle is the side adjacent divided by the hypotenuse. If we assume we are dealing with a circle with a radius of one unit, then the sine and cosine describe the location of our observation on that circle.

If you examine Table 11, you will see that there is a unique combination of values for the sine and cosine of each angle. If the absolute values are identical, then the sine and cosine of the angle differ from all other angles with the same absolute value because of their signs. An example is given below.

ANGLE	SIN	COS
45	.71	.71
135	.71	-.71
225	-.71	-.71
315	-.71	.71

When we take the mean of the sine and the mean of the cosine we are calculating the mean sine of a triangle, where the mean sine is the side opposite and the mean cosine is sine adjacent to the mean angle.

When we calculate r we are actually calculating the hypotenuse (examine how W is calculated and recall the Pythagorean theorem).

By dividing the mean sine and the mean cosine by r we are calculating the sine and cosine of the mean angle. The mean hypotenuse (r) projects from the center of our circle, and since we have assumed a circle with a radius of 1.0, r can never be greater than 1.0.

However, r can vary from 0.0 to 1.0 and still give us the same mean angle. The length of w is therefore an measure of clumping among the observations. If the vanishing bearings were random, then r would approach 0.0. If all of the observations had exactly the same value, then r would equal 1.0.

Our test for significant clumping, not surprisingly, is a function of r (and the sample size). The larger w is, the less likely the points are random. We should note that

$$w = r \times R$$

hence, the maximum value w can have is N (if all of the observations are identical, then $R = N$). As a consequence, it will be more difficult to reject the null hypothesis when we have small sample sizes.

Table 11. Sines and cosines for angles and times of day.

ANGLE	TIME	SIN	COS	ANGLE	TIME	SIN	COS	ANGLE	TIME	SIN	COS
0	0:00	.00	1.00	120	8:00	.87	-.50	240	16:00	-.87	-.50
5	0:20	.09	.99	125	8:20	.82	-.57	245	16:20	-.91	-.42
10	0:40	.17	.98	130	8:40	.77	-.64	250	16:40	-.94	-.34
15	1:00	.26	.96	135	9:00	.71	-.71	255	17:00	-.96	-.26
20	1:20	.34	.94	140	9:20	.64	-.77	260	17:20	-.98	-.17
25	1:40	.42	.91	145	9:40	.57	-.82	265	17:40	-.99	-.09
30	2:00	.50	.87	150	10:00	.50	-.87	270	18:00	-1.00	.00
35	2:20	.57	.82	155	10:20	.42	-.91	275	18:20	-.99	.09
40	2:40	.64	.77	160	10:40	.34	-.94	280	18:40	-.98	.17
45	3:00	.71	.71	165	11:00	.26	-.96	285	19:00	-.96	.26
50	3:20	.77	.64	170	11:20	.17	-.98	290	19:20	-.94	.34
55	3:40	.82	.57	175	11:40	.09	-.99	295	19:40	-.91	.42
60	4:00	.87	.50	180	12:00	.00	-1.00	300	20:00	-.87	.50
65	4:20	.91	.42	185	12:20	-.09	-.99	305	20:20	-.82	.57
70	4:40	.94	.34	190	12:40	-.17	-.98	310	20:40	-.77	.64
75	5:00	.96	.26	195	13:00	-.26	-.96	315	21:00	-.71	.71
80	5:20	.98	.17	200	13:20	-.34	-.94	320	21:20	-.64	.77
85	5:40	.99	.09	205	13:40	-.42	-.91	325	21:40	-.57	.82
90	6:00	1.00	.00	210	14:00	-.50	-.87	330	22:00	-.50	.87
95	6:20	.99	-.09	215	14:20	-.57	-.82	335	22:20	-.42	.91
100	6:40	.98	-.17	220	14:40	-.64	-.77	340	22:40	-.34	.94
105	7:00	.96	-.26	225	15:00	-.71	-.71	345	23:00	-.26	.96
110	7:20	.94	-.34	230	15:20	-.77	-.64	350	23:20	-.17	.98
115	7:40	.91	-.42	235	15:40	-.82	-.57	355	23:40	-.09	.99
								360	24:00	-.00	1.00

(Modified from trigonometric tables in *Standard Mathematical Tables*, S.M. Selby, ed., The Chemical Rubber Co., Cleveland, 1970.)

EXAMPLE

This data set is from an experimental release of bank swallows at a site on the east shore of Seneca Lake in upstate New York. Bearings are in degrees.

a. Calculate V and W

BEARING	SIN	COS	BEARING	SIN	COS	BEARING	SIN	COS
100	.985	-.174	115	.906	-.423	120	.866	-.500
130	.766	-.643	165	.258	-.966	180	.000	-1.000
180	.000	-1.000	185	-.087	-.996	190	-.174	-.985
205	-.423	-.906	210	-.500	-.866	210	-.500	-.866
225	-.707	-.707	230	-.766	-.643	230	-.766	-.643
235	-.819	-.574	240	-.866	-.500	240	-.866	-.500
240	-.866	-.500	245	-.906	-.423	245	-.906	-.423
245	-.906	-.423	245	-.906	-.423	245	-.906	-.423
265	-.996	-.087	265	-.996	-.087	265	-.996	-.087
270	-1.000	.000	275	-.996	.087	280	-.985	.259

$V =$ sin = -14.06, $W =$ cos = -15.50

b. Calculate R.

$$R = \sqrt{(-14.06)(-14.06) + (-15.50)(-15.50)}$$
$$= 20.93$$

c. Calculate the mean sine and mean cosine, and r, and find the mean angle.

$$\overline{\sin} = -14.06/30$$
$$= -.469$$
$$\overline{\cos} = -15.50/30$$
$$= -.517$$
$$r = 20.93/30$$
$$= .698$$
$$\text{sin mean angle} = -.469/.698$$
$$= -.671$$
$$\text{cos mean angle} = -.517/.698$$
$$= -.740$$
$$\text{mean angle} = 222$$

d. Calculate w.

$$w = (20.93)(20.93)/30$$
$$= 14.60$$

e. Since w is greater than 3.00, we reject the hypothesis that the departure bearings of the swallows were random.

Circular Statistics: The Two-Sample Case

APPROPRIATE USES OF THE ANALYSIS

The Rayleigh test is used whenever we wish to know whether a distribution of angles (or times) is random, or whether it clusters significantly about a mean angle. This two-sample case (or Watson and Williams analysis) involves two random samples of bearings. Our null hypothesis is that the mean bearings for the two samples are equal. The alternative hypothesis is that the means are not equal. This test is the equivalent of (forgive the ambiguity) a circular *t* test.

CALCULATIONS

a. Proceed as for the Rayleigh test, keeping the calculations for each sample separate.

V_1 = the sum of the sines for the first sample
V_2 = the sum of the sines for the second sample

W_1 = the sum of the cosines of the first sample
W_2 = the sum of the cosines of the second sample

$R_1 = [(V_1)^2 + (W_1)^2]$
$R_2 = [(V_2)^2 + (W_2)^2]$

b.

$V_p = V_1 + V_2$
$W_p = W_1 + W_2$
$R_p = (V^2 + W^2)$
$N_p = N_1 + N_2$ = the total sample size

c. Calculate the *F* statistic:

$$F = (N - 2)(R_1 + R_2 - R_p)/(N - R_1 - R_2)$$

d. Compare the calculated *F* with that in Table 4 with 1 numerator and $(N - 2)$ denominator degrees of freedom. If the calculated *F* is larger than the tabulated value, the null hypothesis is rejected, and the two means are different.

RATIONALE

If the two mean angles are truly equivalent, then the vector of the combined sample is equal to the sum of the vectors of the two samples ($R_1 + R_2 = R_p$). If the two angles are not equal, then $R_p > R_1 + R_2$. The difference $(R_1 + R_2 - R_p)$ thus measures the deviation between mean angles. If the difference is zero, the angles are equal; if the difference is large, the angles are not equal. The test statistic is most easily derived using the form given for the *F* statistic in paragraph c above.

EXAMPLE

This hypothetical set of data indicates the times at which two species of flies were trapped on sticky flypaper suspended in the men's room of the Greyhound Bus Station in West Palm Beach, Florida.

Species A 5:00, 5:00, 5:20, 5:20, 5:20, 6:20, 8:40, 11:20, 14:00, 5:20

Species B 19:00, 19:00, 21:40, 23:40, 0:40, 3:20, 3:40, 3:40, 4:20, 6:00

a. Calculate V, W, and R for each sample.

$$V_1 = 6.32 \qquad V_2 = 1.89$$
$$W_1 = 1.55 \qquad W_2 = 5.53$$
$$R_1 = 6.51 \qquad R_2 = 5.84$$

b. Calculate the pooled V_p, W_p, and R_p.

$$V_p = 8.27$$
$$W_p = 3.98$$
$$R_p = 9.12$$

c. Calculate the F value.

$$F_{(1,17)} = (19 - 2)(6.51 + 5.84 - 9.12)/(19 - 6.51 - 5.84) = 8.26$$

d. Compare the calculated F (8.26) with that obtained from Table 4 for 1 and 17 degrees of freedom. Since the calculated value clearly exceeds the tabular value (4.45), we reject the null hypothesis and conclude that the two groups had different mean activity times.

Analysis of Skewness and Kurtosis

APPROPRIATE USES OF THE ANALYSIS

Many of the variables that we are interested in studying have approximately normal frequency distributions. Others, however, have recognizably non-normal distributions. While there are many different ways in which a distribution can differ from normality, two are of special statistical interest: skewness and kurtosis. A **skewed distribution** is one in which one of the tails is drawn out and the other is truncated. Thus skew to the right means that the right hand tail is drawn out, while skew to the left means that the left hand tail is extended. **Kurtosis** refers to the height or "peakedness" of the distribution. A **leptokurtic distribution** has more observations clustered around the mean and more in the tails than does a normal curve. A **platykurtic distribution** has more observations in the regions between the mean and the tails than does a normal curve.

There may be many occasions when we want to know if a frequency distribution is skewed or has appreciable kurtosis. We might, for example, measure elytra lengths of male beetles captured in copula and ask whether the frequency distribution of lengths is skewed to the left, which would suggest that selection favors larger elytra and that there is an upper limit to elytra size. We might, for example, measure distances moved by dispersing animals and ask whether the frequency distribution is leptokurtic, which would suggest that some animals moved very little while others moved great distances. These types of questions can be answered by evaluating skewness and kurtosis through application of the statistics K_1 and K_2.

CALCULATIONS

a. Calculate the following:

   Total sample size = N

   Sum of all the observations = ΣY_i

   Sum of all the squared observations = ΣY_i^2

   Sum of all the cubed observations = ΣY_i^3

   Sum of all the observations raised to the fourth power = ΣY_i^4

b. Calculate the following statistics:

   1. Sample mean = $\Sigma Y_i / N = \overline{Y}$
   2. Sample variance = $[\Sigma Y_i^2 - (\Sigma Y_i)^2/N]/(N - 1) = S^2$
   3. $K_1 = [(1/N)(\Sigma Y_i^3) - (3\overline{Y}/N)(\Sigma Y_i^2) + 2\overline{Y}^3]/S^3$
   4. $K_2 = [(1/N)(\Sigma Y_i^4) - (4\overline{Y}/N)(\Sigma Y_i^3) + (6Y^2/N)(\Sigma Y_i^2) - 3\overline{Y}^4]/S^4$

c. In a normal distribution, the expected value of K_1 is zero and of K_2 is 3. The significance of K_1 can be evaluated through calculation of a t statistic:

$$t = (K_1 - 0)/(24/N)$$

The null hypothesis is that the calculated K_1 is not different from zero. The null hypothesis is rejected if $t > 1.96$, and we conclude that the distribution is skewed to the left if K_1 is negative and skewed to the right if K_1 is positive.

The significance of K_2 can be tested in a similar manner, but in this case:

$$t = (K_2 - 3)/(24/N)$$

If the quantity $(K_2 - 3)$ is less than -1.96, the distribution is platykurtic. If $(K_2 - 3)$ is greater than 1.96, the distribution is leptokurtic.

d. Note that all calculations are much easier if you begin by organizing the data into a frequency distribution (as has been done in the example).

EXAMPLE

The following data were collected by marking female soldier beetles *(Chauliognathus pennsylvanicus)* and recapturing marked individuals after 24 hours. For convenience in this example, movements have been summarized in 10-m classes and class midpoints are used in the calculations.

DISTANCE (Y_i)	FREQUENCY (f_i)	f_iY_i	$f_iY_i^2$	$f_iY_i^3$	$f_iY_i^4$
5	60	300	1500	7500	37500
15	6	90	1350	20250	303750
25	4	100	2500	62500	1562500
35	0	0	0	0	0
45	5	225	10125	455625	20503125
55	2	110	6050	332750	18301250
65	1	65	4225	274625	17850625

a.

$$N = \Sigma f_i = 78$$

$$\Sigma(Y_i) = \Sigma(f_iY_i) = 890$$

$$\Sigma(Y_i)^2 = \Sigma(f_i)(Y_i)^2 = 25{,}750$$

$$\Sigma(Y_i)^3 = \Sigma(f_i)(Y_i)^3 = 1{,}153{,}250$$

$$\Sigma(Y_i)^4 = \Sigma(f_i)(Y_i)^4 = 58{,}558{,}750$$

b.

$$\overline{Y} = 11.41$$

$$S^2 = 202.53$$

$$S = 14.23$$

$$K_1 = 2.24$$

$$K_2 = 6.90$$

c. To test the null hypothesis that $K_1 = 0$ (i.e., that there is no skew),

$$t = 2.24/0.28 = 8$$

which exceeds 1.96; thus we reject the null hypothesis.

To test the null hypothesis that $K_2 = 3$ (i.e., that there is no kurtosis),

$$t = (6.90 - 3)/.55 = 7.09$$

which exceeds 1.96; thus we reject the null hypothesis.

d. In this case, the finding that the distribution is skewed to the right is not surprising, since the distribution is clearly asymmetrical. The finding of interest involves K_2, which is large enough to support the conclusion that the observed distribution of movements was leptokurtic, suggesting that dispersal did not generate a half-normal curve and that distances moved by individuals during the study were not equivalent.

Table 12. Random Numbers.

10525	96327	58276	33115	70286	42458	31039	41809	94505	25156
68635	80772	16109	63609	54329	01254	87210	58881	24038	81806
46973	19086	68684	40831	05318	37826	17267	84079	39836	47729
56140	72434	53600	88708	41312	28466	83608	13802	01924	95194
75858	75864	43053	58617	04277	58991	95074	61374	50800	61299
13519	12673	70747	98668	02228	48331	03969	15976	80631	78384
83429	26663	45122	64008	28314	41250	34720	06669	97213	50160
78634	56661	95791	27516	78746	04590	78560	68351	70066	45407
32203	02680	25055	75259	16413	86402	13591	31556	13054	06129
89832	58101	61367	96095	52087	22338	34621	82955	52888	92035
40656	67238	36630	99785	40075	98691	02139	68427	69898	82270
78137	92672	34193	23002	10207	06449	44802	35590	20448	87962
87908	17270	79427	07600	83012	45736	74821	37660	75358	27370
84438	06461	69756	60527	39990	65023	00602	45511	33727	93999
43410	16542	66013	29753	45826	44475	46036	39811	96516	56059
01672	89402	92000	55587	44434	57820	10802	80119	09272	04255
21635	87019	14130	81764	75705	72100	31060	72800	08367	63060
42015	73718	98014	35667	84835	69697	24910	84649	30132	13329
61761	14371	63884	12066	21104	01992	31988	74094	67252	73100
29071	57350	87718	19754	47980	09046	81589	36851	81801	08495
32361	11630	19934	75437	11435	64664	48666	87903	09297	71549
15055	52402	18552	54277	38330	31233	82104	26687	13148	86890
61089	14581	18673	31441	61758	33693	14633	57359	73853	34107
22982	58576	85164	74046	95211	87032	88754	61010	22361	97703
89370	17701	20423	67866	02459	36661	58058	41710	51928	81497
13859	61606	24095	88184	34997	96066	48310	93257	67881	09360
52087	35095	26618	49158	33950	91091	87673	58503	59631	59147
78864	56702	53470	91072	18304	45244	40452	78133	04341	01714
82264	86432	21994	89862	06345	65386	03718	50774	36602	41881
56542	96983	77681	98277	42170	12195	51211	51302	43502	83230
40761	07890	41773	53058	24155	64763	33872	79235	69458	63602
74988	66915	67079	07501	57249	13711	71399	34070	40192	33174
70954	88086	63856	73798	75973	42196	94192	39460	40223	71986
19033	80750	62658	84507	09995	79180	20056	63078	03243	68801
78429	29016	26003	07545	89160	34347	75054	04075	99770	08164
22437	05963	55746	74066	09849	49026	92876	44983	31428	37542
35665	31991	68030	03609	09536	22510	52087	89945	66257	11413
85072	11687	32690	35887	21430	98144	35655	36153	74399	34190
11723	08578	98986	03586	06065	18544	22110	50359	48564	48840
15109	42156	54541	05561	81720	84342	42723	47347	18625	43089
26933	16535	89350	64794	00542	75812	64073	92220	94734	70074
96211	15135	61750	80494	42249	21773	80382	73408	98328	05323
32585	20747	88212	08711	22282	25118	24990	18233	65403	86948
79713	45366	03845	30845	36288	79376	35345	47948	31422	10900
15610	79168	65464	94413	87807	35646	30882	94088	32245	78192
01816	92999	94126	35466	70034	04314	88172	23036	80449	15942
28272	53755	78984	17012	05535	00896	22708	40655	01408	19119
25770	37031	89349	42837	44795	70408	11708	73886	38519	84871
42855	46405	66829	31090	05464	49620	17309	51164	61946	36307
09064	73745	94417	57004	89182	51578	94523	28543	40245	57599

Summary of Labs by Location and Season

	Location		Season		
	Lab	Field	Autumn	Winter	Spring
Mechanistic Approaches...	X		X	X	X
Goal Orientation...		X	X	X	X
Color Perception...		X	X		X
Perception in Bats...	X		X		X
Prey Location...		X	X	X	(X)
Releasers and Egg Laying...	X				X
Food Value...		X	X	X	(X)
Foraging Patterns...	X		X	X	X
Seed Predation...		X	X	X	(X)
Avian Foraging...		X	X	X	(X)
Search Images...	X		X	X	X
Flower Choice...		X	X		X
Predation Efficiency...	X		X	X	X
Resource Partitioning..		X	X		X
Dispersal and Dispersion...		X	X		X[1]
Fish Schools...	X		X	X	X
Bag Construction...		X	X	(X)	
Spacing of Eggs...		X			X
Human Groups...		X	X	X	X
Oviposition by ...	X		X	X	X
Assortative Mating...	X[1]	X	X	X[1]	X[1]
Sex ratios...	X		X	X	X
Oviposition Site...	X	X	X		X
Male Dominance...		X	X		X[1]
Mate Preferences...	X		X	X	X
Rare Male...	X		X	X	X
Male-Male Competition...	X		X	X	X
TOTAL	14	15	25	18	26

Parentheses indicate that this is not the most appropriate season.

[1]Can be done with other species